THE HANDBOOK
ON
OLEANDERS

by

Richard & Mary Helen Eggenberger

First Edition

Mrs. F. Roeding

Richard & Mary Helen Eggenberger

THE · HANDBOOK · ON
OLEANDERS

with photographs by the authors

Tropical Plant Specialists

Cleveland, Georgia

First edition published in 1996 by
Tropical Plant Specialists
Cleveland, Georgia

Unless otherwise noted, all
Photographs
by
Richard and Mary Helen Eggenberger

Front Cover
Apple Blossom
Back Cover
Sorrento

Library of Congress Catalog Number
96-60700
ISBN 0-9643224-0-4

Scarlet Beauty

Kewpie

DEDICATION

IN MEMORIAM

Maureen Elizabeth "Kewpie" Gaido
1916-1995

Friend and champion of the Oleander.
Through her love, enthusiasm and vivacious
personality, thousands of gardeners have
been introduced to this magnificent family
of plants. Wherever she traveled she spoke
eloquently of utilizing the beauty of
oleanders to enhance our landscapes.
All who love flowers owe her a debt of
gratitude for her indefatigable efforts
in promoting oleanders world-wide.

CONTENTS

Under the branches
Of the cherry-trees in bloom,
None are strangers there.
Issa

Acknowledgements

We who love plants and flowers are especially blessed for we belong to one of the world's largest and most beneficent families, those who garden. To work with the earth, to witness her infinite manifestations of beauty, birth and fruition, to share with friends in every geographical climate plants, seeds and horticultural experiences, to awaken each day to new exploration within and without, to plant a seed or a tree, to marvel at the song of a bird or the fragrance of a blossom, and to witness beauty beyond description imbues one with an eternal sense of gratitude and awe. In the words of India's great sage and poet Sri Aurobindo:

"All's miracle here and can by miracle change."

(from Savitri)

In recent times the oleander has gained greatly in popularity in the United States due in large part to efforts of growers and hybridizers, including champions of the oleander such as Kewpie Gaido, Clarence Pleasants, Bob Newding, Elizabeth Head and many other members of the International Oleander Society. During the past few years we have had the opportunity to correspond with many of these dedicated people throughout the United States and Europe who have made invaluable contributions on behalf of the oleander and have been fortunate in being able to personally interview and tape record hours of their reminiscences. This list of those who have assisted us and answered seemingly endless questions is lengthy indeed and we express our personal gratitude to each and every one.

Our special thanks is due to the following people and institutions who have been instrumental in helping us realize our goal of a comprehensive work on oleanders that would be of interest to both amateur and professional gardeners, plant collectors, nurserymen and hybridizers.

To the late Maureen Elizabeth "Kewpie" Gaido, founder of the International Oleander Society, whose enthusiasm, gracious hospitality and personal encouragement was the primary impetus for this book. We shall always remember our visits to her in Galveston, her love of flowers, especially the oleander, her welcoming smile and lasting friendship.

To Elizabeth S. Head, Historian and Corresponding Secretary of the International Oleander Society, and Founder and Editor of *Nerium News*, for sharing some of the society's impressive collection of data on oleanders as well as her knowledge of oleander cultivars in Galveston, for sending us past issues of *Nerium News* for reference, and for the time she devoted to assisting us, not the least of which was driving me around Galveston to seek special cultivars of oleanders to photograph.

To Bob Newding, whose assistance in the technical sections of this book has been invaluable. His selflessness in sharing his technical expertise and encyclopedic knowledge of oleander varieties and culture through hours of recorded interviews, the time he has devoted to showing us oleander plantings throughout Galveston, his illustrations of particular aspects of pruning, planting and design, and his cheerful goodwill and friendship exemplify the deep and lasting camaraderie amongst all who love horticulture.

To Ted Turner, Sr. and Ted Turner, Jr., for their generosity and hospitality, a wild and wonderful sense of humor, and particularly for sharing their love of the nursery industry, their delight in growing things, and their unparalleled experiences with oleanders and the magnificent cultivars that are the result of years of dedicated labor.

To Frank J.J. Pagen, plant taxonomist and author of the finest technical work on the genus, *Oleanders, Nerium L. and the oleander cultivars*, for the extensive research he conducted on the genus, for his kind and generous permission to quote extensively from his book and reproduce his excellent line drawings, and for sharing his vast experience with European cultivars.

To Dr. Barry Comeaux, for sharing his experiences as President of the International Oleander Society and the many scientific research projects initiated during his tenure.

To Professor Octavia Hall, for so kindly sharing her experiences in preparing herbarium specimens of oleanders and the invaluable treasure of her illuminating notes from conversations with Clarence Pleasants, providing the most thorough history of his life and accomplishments, and for permission to quote from her writings.

To George Sealy III, for sharing his reminiscences of the Sealy family, especially his father's generosity in supplying hundreds of thousands of oleanders to the public free of charge, and for his priceless and humorous anecdotes.

To Lane Taylor Sealy, who at seven years of age was one of the first "oleander rustlers" in Texas, for recounting the saga of the Sealy family and glimpses of Galveston's early history, and for regaling us with stories of his life with his father George Sealy, Jr. among the oleanders.

To John Kriegel, Director of Gardens and Gary Outenreath, Exhibits Manager, both of Moody Gardens in Galveston, Texas, who have helped amass one of the world's great oleander collections at Moody Gardens, for their kindness and willingness to share their vast horticultural expertise.

To the Moody Foundation, for their interest in highlighting oleander plantings in addition to many other wonderful experiences offered at Moody Gardens.

To Dr. Darrell MacDonald, cultural plant geographer by self-definition, bio-geographer by discipline, and ethnobotanist by many other standards, for sharing his PhD dissertation research which focused on the transformation of Galveston from a barrier island into an urban complex of exotic plantings and native plants.

To Dr. Jerry Parsons, Professor and Extension Horticulturist of Texas A&M University Agricultural Extension Service at San Antonio, for his illuminating and humorous accounts of his experiences with oleanders.

To Dr. Wayne McKay, Research Horticulturist and Director of Texas A&M University Agricultural Extension Service at El Paso, for sharing his research into the development of increased cold tolerance using irradiated oleander seeds.

To Dr. David Morgan, for permission to quote from the article, "How to Register Patents, Trademarks," in the July 1989 issue of *Nursery Manager*.

To the Sri Aurobindo Ashram, Pondicherry, India, for permission to quote from the works of The Mother and Sri Aurobindo.

To the Bishop Museum in Honolulu, Hawaii, for permission to quote from Marie C. Neal's book, *In Gardens of Hawaii*.

To the Monrovia Nursery Company and Roger Duer, for supplying us with the background on the Monrovia hybrids including original photographs.

To Gopalaswamy Parthasarathy, for permission to quote from K.S. Gopalaswamiengar's book, *Complete Gardening in India*.

And finally, to a special friend and pioneer in the field of oleanders, the late Clarence Grant Pleasants, author of the first book on the genus entitled *Galveston, The Oleander City*, for his energy and devotion to promoting his beloved plant, his inspiration for the International Oleander Society to be founded in Galveston thereby assuring that it would forever be known as The Oleander City, his selfless sharing of plant material with botanical gardens and institutions, and his personal kindness to us in hours of recorded interviews relating the extraordinary history of his experiences with oleander cultivars, their culture, habits and special attributes. Clarence passed away December 27, 1995, as this book was in preparation.

Gratitude is the memory of the heart.
Anonymous

Mrs. Burton

If thou wouldst attain to thy highest, go look upon a flower;
what that does will-lessly, do thou willingly.
Friedrich von Schiller

Introduction

What a great moment in time to be on earth witnessing such vast changes on all levels of human endeavor, experiencing almost daily technological breakthroughs that promise longer and healthier life, the exchange of information on unprecedented levels and the nascent possibilities of peace on the global horizon.

One of the frontiers of our age where technology, working hand in hand with Mother Nature, has before it a unique opportunity to forge exciting new possibilities is in the world of plants. Only a few generations ago horticulturists and growers were stretching the envelope with the idea of crossing species, or hybridizing, and most of the monumental task of identifying and naming plant genera and species was carried out worldwide by a handful of adventurous, dedicated and indomitable plantsmen. Today, even in the high-tech culture that permeates our daily lives — from microwave cooking to virtual reality systems that may soon allow us to experience a plant from the molecules of its vascular system — it is still somewhat amazing to realize that botanists are not only enhancing existing species by selective breeding, but are literally creating new species through the technique of gene splicing. And yet, on a warm spring day when every plant and flower seems bursting with its own perfect beauty, it seems we could live a thousand years yet barely begin to experience and appreciate what Mother Nature has already placed at our doorstep. Perhaps through this book we can open another small door.

During the months of May and June we may immerse ourselves in the colorful and often fragrant world of oleanders. One can find major oleander plantings in the United States in Zones 9 and 10 and the warmer parts of Zone 8, especially the coastal regions of South Carolina and Georgia, and in the southernmost areas of the Gulf Coast states. Spring through early summer is an ideal time to travel a few of the thousands of miles of freeways and highways in California that are graced with these exceptionally durable plants whose toughness and long-flowering qualities are legendary. This is also the best season to visit Galveston and Corpus Christi in Texas, to view thousands of stunning, floriferous shrubs laden with colorful blooms, or to travel to Disney World to witness at their peak the masses of oleanders, especially the extraordinary tree forms, that have inspired so many visitors with their cheerful colors.

1

These days it seems we find references to oleanders everywhere: *Nerium News*, the newsletter of the International Oleander Society, tells of Oleander Avenue in Niles, Illinois, and Mr. Donald Becker who grows oleanders in pots and distributes them to other residents of Oleander Avenue hoping one day the street will be filled with oleanders in bloom. We also visited with a CEO of a major real estate development firm, an avid golfer, who spoke of having recently played the "Oleander Course" at Jekyll Island, Georgia. A current issue of *American Record Guide* features an article on new music in Russia by Harlow Robinson where we learn about the music festival in Sochi, Russia, a town with a population of 400,000 in the Black Sea resort area that is a subtropical paradise "with its rocky azure beaches and blooming oleanders. . . ."

Oleanders are grown extensively in warm coastal areas around the world. Anyone who has ever taken a holiday in the Mediterranean could not fail to be impressed by the oleanders that are in bloom everywhere, especially on the Greek Islands. The oleander has been a beloved plant for centuries in Europe, the Middle East and China, and it is extensively cultivated in public and private gardens in Africa and Australia. We have been inspired by magnificent plantings in Mexico, on tropical islands, and in gardens of India and Southeast Asia. In fact, in our years in India we rarely saw a garden that did not have a planting of oleanders. In *Complete Gardening In India*, K.S. Gopalaswamiengar writes: "The Oleanders are some of the most delightful of fine flowering shrubs, which no Indian garden is without."

Our fascination with oleanders and their importance in the landscape began nearly thirty years ago when we started building the Matrimandir Gardens in Auroville in southern India. We traveled the length and breadth of the Indian sub-continent studying and collecting thousands of flowering tropicals to introduce into the gardens. Our journeys took us to many other tropical areas as well, searching for new plants in Thailand, Singapore, Sri Lanka, Hawaii, Mexico, Japan, the Caribbean Islands and the continental United States. Interestingly, our lasting memory of ole-anders in Asia is of the strong and delightfully sweet scent of the flowers. Although oleanders were easily grown and propagated in southern India, requiring little care even in that challenging climate, there were only a few varieties available. It was only when we returned to the United States in 1981 and moved to Houston, Texas, where we had the opportunity to visit Galveston, that we began to realize the great wealth of flower colors, the vibrant shades which ranged from red through pink to yellow and white, and the variety in sizes and shapes that are now available. Subsequent visits to California reminded us of their excellence as mass plantings to soothe the highway traveler along mile after mile of freeways. Visits to Corpus Christi, Texas, introduced us to other spectacular hybrids, from compact, petite and dwarf forms eminently suitable for container culture to small trees offering months of splen-did bloom. The delicate beauty of its flowers coupled with the oleander's ability to withstand heat, drought, high winds, saline conditions, freeway traffic and air pol-lution and still continue to bloom for months at a time was all the impetus we needed to write about this very special genus, *Nerium*. We hope to bring this tough, adapt-able and beautiful plant to the attention of the thousands of gardeners who have followed our efforts to introduce unique and colorful tropicals and subtropicals to

American gardens, and to encourage people everywhere to grow some of the new, free-blooming cultivars as well as some of the great varieties that have been developed during the past 100 years.

And so we begin a journey to explore another special flowering plant, the oleander, to learn that it has been grown and loved for centuries and is undoubtedly one of the plants spoken of in the Bible and other ancient religious texts. In India the flowers are thought by many to hold special spiritual significances. The following is from the book *Flowers and Their Messages* by The Mother, a French woman by birth, who spent her life in India and became revered for her work and teachings.

On the Significance of Flowers

Question to the Mother: *"How do you give a significance to a flower?"*
Answer: *"By entering into contact with it and giving a more or less precise meaning to what I feel . . . by entering into contact with the nature of the flower, its inner truth; then one knows what it represents."*

Single pink with light yellow throat	*Sweetness of Thought Turned Exclusively towards the Divine*
Double pink	*Surrender of All Falsehood*
Single magenta or deep red	*Changing of Wrong Movements into Right*
Single, pale pink	*Contemplation of the Divine*
Single white	*Quiet Mind*
Double white	*Perfect Quietness in the Mind*
Single white with elongated petals	*Quietness Established in the Mind*

Flowers speak to us when we know how to listen to them,
– it is a subtle and fragrant language.
The Mother

I. Lovenberg

Flowers are the moments representation
Of things that are in themselves eternal.
Sri Aurobindo

Oleanders and the Apocynaceae Family

Oleanders are members of one of the most colorful groups of plants in the horticultural kingdom, the Apocynaceae or Dogbane Family. The family was named by A.L. de Jussieu in 1789. Accounts as to the number of genera are varied with *Hortus III* of the L.H. Bailey Hortorium citing about 130 and the Royal Horticultural Society's *Dictionary of Gardening* listing 215. The more than 2000 species within the family are comprised of trees, shrubs, vines, herbs, succulents and groundcovers, mostly of tropical origin, with many being the source of drugs (glycosides) to treat heart disease and rubber which is commercially obtained from others. Some of the genera, popular for their ornamental value, have seen an explosion of interest from hybridizers and the gardening public in the past twenty years and have led to many exciting new varieties.

Since the late 1980's we have seen a burgeoning interest in flowering tropicals. Plants such as *Mandevilla*, unheard of by the greater gardening public a decade ago, are now regularly seen at nurseries everywhere and enjoyed in gardens throughout the country. New varieties of *Allamanda*, *Bougainvillea*, ginger and *Plumeria* appear every year.

The genus *Nerium*, or oleander, is certainly near the top of the list in popularity. Other members of the family well known to gardeners in the United States are: *Adenium obesum*, Desert Rose; *Allamanda spp.*; *Carissa*, Natal Plum; *Catharanthus*, commonly known as Vinca or Periwinkle; *Mandevilla spp.*; *Pachypodium*; *Tabernaemontana*, known in much of the United States as Carnation of India and elsewhere as Crepe Jasmine; *Thevetia*, Yellow Oleander; *Trachelospermum*, Star Jasmine or Confederate Jasmine, and its groundcover relative the Asiatic Jasmine; and *Vinca major* and *Vinca minor*, also known as Periwinkle, those ubiquitous and attractive groundcovers for shade and partial sun available in blue, white or lavender-purple flowers.

Listed below are brief descriptions of a few of the family members that are allied to the colorful oleanders. (See photos, pgs. 8 and 9)

Adenium obesum — Desert Rose. The genus *Adenium* contains about four species native to tropical Africa. *A. obesum*, the Desert Rose, is a slow-growing plant that lends itself admirably to bonsai culture with its swollen caudex and large, red to pink flowers set against dark green leaves. This is a very drought resistant plant that stores water in its stem and is relatively free of pests and disease. In recent years many new cultivars with flowers in shades of cream, rose-purple and lavender have been introduced. We have seen a report in a recent issue of *Nerium News* from our longtime friend Mr. R. Haresh of Madurai, India, indicating that hybridizers there have crossed the *Adenium* with *Nerium oleander*. We can't wait to see the results as they are certain to be unique! F.J.J. Pagen, in *Oleanders, Nerium l. and the oleander cultivars*, states that *Adenium* is the most closely related genus to *Nerium* and that *Adenium* species may be grafted on *Nerium oleander* rootstock to produce profusely flowering plants.

Allamanda — A genus of approximately twelve species, all native to tropical America. The showy, often very large flowers range from cream to yellow to purple and are borne in profusion over a long blooming season. Many new cultivars have been introduced during the 80's and 90's. Some of the new hybrids are notable for a number of characteristics such as dwarf forms suitable for groundcovers and container culture, and new colors in shades varying from cream to rose-purple.

Carissa — A genus native to the Old World Tropics containing thirty-five species of small evergreen shrubs and trees, mostly with spines. The most popular members make handsome hedges or ground covers and produce an abundance of very fragrant, white to pink flowers. Some species produce edible fruits that may be made into an excellent jam. Pictured on Pg. 9 is one of the compact forms *Carissa grandiflora* 'Compacta', a low-growing ground cover or small hedge that flowers freely throughout spring and summer.

Catharanthus — About five species of erect annual or perennial herbs native to the Old World Tropics belong to this genus. The most popular form is *Catharanthus roseus*, known to most gardeners around the world as Vinca (Annual), Periwinkle, Madagascar Periwinkle or Old Maid. In the past ten years plant breeders have created some of the most beautiful hybrids imaginable with much larger flowers, greater disease resistance, more compact forms and a wealth of new colors.

Mandevilla — All *Mandevillas* are native to tropical America, especially Argentina, Bolivia and Brazil. This is a large genus comprising nearly one hundred species, usually vines and scandent shrubs. Extensive breeding work is in progress in the mid 1990's and some of the new hybrids are truly breathtaking. We have seen recent crosses of M. 'Alice du Pont' with deep red flowers, pale pink and white flowers, and a pure white hybrid, all with blossoms as large or larger than the original cultivar!

Mandevilla x amabilis 'Alice du Pont' — A magnificent vine of moderate growth producing large, bright pink flowers (up to 4 inches across) set against medium green leaves. An excellent plant for containers as it flowers from spring to fall in most areas.

6

Mandevilla boliviensis — White Dipladenia. A species that has become very popular in recent years and is propagated by major growers across the United States. This is a small vine that is easily grown in containers and one we've seen flower in 4 inch pots, in bloom from spring through fall.

Mandevilla sanderi 'Red Riding Hood' — An excellent variety for container culture, especially hanging baskets, as it tends to spill rather than climb. Flowers are smaller than the varieties mentioned above but are unique for their rich, deep pink color. Plants are especially attractive for the small, glossy, deep green leaves that have a bronzy tinge when young.

Mandevilla sanderi 'Rosea' — Brazilian Jasmine. A delightful small vine with glossy, dark green leaves that are bronze beneath and in the new growth. Flowers are trumpet-shaped in a delicate, clear rose-pink with a yellow throat. Plants bloom on and off throughout the growing season and sporadically all year in warm climates or greenhouses.

Plumeria — Frangipani, Temple Tree, Hawaiian Lei Flower. Since plumerias were the subject of our first book you can imagine that we are partial to them. Available in a rainbow of colors with large, often intensely fragrant flowers, new dwarf hybrids, and a flowering season that extends from spring until fall, plumerias are an excellent choice as companions for oleanders. Since they are not nearly as hardy, however, they must be given full winter protection except in tropical climates. The following three plumerias are representative of hundreds of cultivars and species grown around the world:

Plumeria 'Aztec Gold' — Large, buttercup yellow flowers with a faint pink band on the reverse and a heady, strong fragrance reminiscent of fresh peaches.

Plumeria 'Kimi Moragne' — Very large, sweetly fragrant, intense rose-pink flowers shading to lighter pink at the extreme edge of each petal and a golden-yellow center. Clusters are very dense with many blooms open at a time. One of many extraordinary hybrids created by the late Bill Moragne in Hawaii.

Plumeria 'Singapore' (*P. obtusa*) — This species bears pure white, perfectly shaped flowers with small yellow centers and the typical, exotic, Frangipani fragrance. Unlike varieties of *P. rubra*, plants are evergreen and have thick, leathery, glossy, dark green leaves.

Tabernaemontana divaricata — Crepe Jasmine, Carnation of India. A very handsome shrub of moderate growth with soft white flowers set against glossy, dark green leaves. Select cultivars are available with single, semi-double and fully double flowers. Recently a very compact dwarf form was introduced.

Thevetia peruviana (T. neriifolia) — Yellow Oleander, Be-Still Tree. A colorful member of the Apocynaceae family often used as a hedge or screen in tropical climates. The term Yellow Oleander implies that it was formerly believed to be a

Family Members

Mandevilla x amabilis
'Alice duPont'

Mandevilla boliviensis
White Dipladenia

Urichites lutea
Yellow Mandevilla

Allmanda Williamsii
'Stansill's Double'

Mandevilla sanderi
'Red Riding Hood'

Mandevilla sanderi
'Rosea'

Thevetia peruviana
Yellow Oleander

Allamanda cathartica
var. 'Hendersonii'

Adenium obesum
Desert Rose

8

Family Members

Catharanthus roseus
Madagascar
Periwinkle

Tabernaemontana
divaricata
Crepe Jasmine

Plumeria
rubra acutifolia
'Kimi Moragne'

Carissa grandiflora
'Compacta'

Allamanda cathartica
'Cherries Jubilee'

Plumeria obtusa
'Singapore'

Trachelospermum
jasminoides
Star Jasmine

Plumeria
rubra acutifolia
'Aztec Gold'

variety of oleander. Forms with white, yellow and orange flowers exist as well as a rare, large-flowered species **T. ycotli** with buttery yellow blooms. All parts of this plant are extremely toxic and we have known the fruits to be chewed by people in southern India to commit suicide.

Trachelospermum jasminoides — Star Jasmine, Confederate Jasmine. A favorite ground cover plant in southern California, especially on slopes, and highly valued everywhere for its sweet fragrance that can perfume an entire garden. In cooler areas it is a superb container specimen, especially striking set against a trellis with its white flowers appearing like stars against the dark green foliage. Virtually evergreen in Zones 9 and 10, it will withstand below freezing temperatures for short periods. A yellow flowered form **T. jasminoides 'Mandianum'** is also available.

Urichites lutea — Yellow Mandevilla. A small flowering vine only recently introduced to American gardeners. The glowing golden-yellow, trumpet-shaped flowers are set against attractive, glossy, medium-green leaves.

<u>Vinca</u> — A genus consisting of approximately twelve species, all native to the Old World. The two most popular for ground covers and window boxes are listed below:

> **V. major** — Greater Periwinkle, Blue-Buttons. An excellent ground cover for shady, moist areas, this hardy plant is grown throughout much of the United States. A few unusual varieties are available including a variegated form 'Variegata' (or 'Elegantissima'). Flower color varies from blue to deep purple.

> **V. minor** — Common Periwinkle, Lesser Periwinkle, Myrtle, Running Myrtle. Generally hardy to Zone 3 and further north if heavily mulched or protected by snow. Growth habit is very prostrate with plant height from 3 to 6 inches. Can take full sun in more northerly zones but prefers some shade from Zone 8 south. One of the best ground covers for its compactness, ease of culture and lovely flowers that range from white to violet-pink, blue and purple, with the most frequently seen shade a lovely lilac-blue.

Quiz

If the question were asked, "What plant can you name that has <u>all</u> of the following attributes?"...

1. Drought tolerance

2. Insect resistance

3. Disease resistance

4. Wind tolerance

5. Salt tolerance

6. Smog tolerance

7. Fast growing

8. Very easy to grow

9. Ideal container plant

10. Superb windbreak

11. Many dwarf forms

12. Highly fragrant forms

13. Able to withstand intense heat, even that reflected off concrete and asphalt

14. Able to withstand automobile pollution and frequently planted on freeways

15. Able to grow in widely divergent soil conditions from sand to clay and gravel

16. Grows equally well in the desert and at the seaside

17. Needs little fertilizer but responds well to fertilizing

18. Excellent choice for hedges and privacy screens

19. Tolerates light freezes and although sometimes killed to the ground in temperatures below 20°, will regenerate in the spring

20. As a landscape subject makes an ideal background plant (large shrub forms), middle ground specimen (intermediate forms), and foreground subject (dwarf forms)

21. Evergreen in warm climates

22. Available in a wide range of colors

23. Flowers abundantly from spring into fall

24. Cultivars available with single or double flowers

25. Ideal foundation plant against a wall (See photo Pg. 104), border plant (dwarf forms), or as a foil or complement in architectural settings

26. Produces large clusters of extraordinarily delicate and lovely blooms

. . . Your answer might well be, "None, such a plant doesn't exist. It's a figment of your imagination!" Fortunately for gardeners around the world such a plant does exist and it's the oleander, a plant that for many rivals in beauty the camellia of northern gardens. Few plants offer so much beauty with so little care.

Flowers . . . have a mysterious and subtle
influence upon the feelings, not unlike some
strains of music. They relax the tenseness of
the mind. They dissolve its rigor.
Henry Ward Beecher,
Eyes and Ears

Barbara Bush

2

. . . It is strange that a little mud
Should echo with sounds, syllables and letters
Should rise up and call a mountain Popocatapetl,
And a green-leafed wood Oleander. . .
W. J. Turner,
Talking With Soldiers

Nomenclature

Origin of the Latin Name

Joseph Pitton de Tournefort, a botanist of the late 17th century (1656-1708), established the genus *Nerion* in 1700; the name was later Latinized by Linnaeus who, in 1737, changed it to *Nerium* as it is known today.

The best description we have for the origin of the genus is from the Dutch taxonomist F.J.J. Pagen. In Part One of his book *Oleanders: Nerium L. and the oleander cultivars*, he tells us that the name *Nerium* is derived from the Greek word *nerion* and was used by Dioscorides to indicate the oleander. Supposedly the name refers to the Greek sea-god Nereus and his fifty daughters, the Nereides. The ancient Greek maintained holy forests planted exclusively with oleanders and garnished altars to honour the Nereides who were considered to be infallible guides. Others suggest that *nerion* is derived from the Greek *neros*, meaning "moist," a reference to areas of available water where oleanders were most likely found in an arid countryside. According to Pliny, the Greeks also knew oleanders by the names *Rhododendros* and *Rhododaphne*.

The specific epithet *oleander* is derived from two languages; *olea*, in Latin, meaning olive tree (the leaves of the olive and the oleander are similar and were thus thought to be related) and *dendron*, the Greek term for "tree" or "shrub." Linnaeus applied the binomial *Nerium oleander* in his *Species Plantarum* in 1753.

Classification of Species

Although the literature is replete with references to numerous species, authorities today agree that the genus *Nerium* contains only one species, *Nerium oleander* L., with *N. indicum* Mill. and *N. odorum* Ait. being superseded names.

The contemporary authority A.J.M. Leeuwenberg published a series of revisions of the Apocynaceae in 1984. Pagen states that from comparative studies of several hundreds of herbarium specimens and many living plants in gardens, Leeuwenberg concludes that they all should be considered members of the single species *Nerium oleander* L.

Let Woodward, house of Woodward,
rejoice with Nerium the Rose-laurel, ...
Christopher Smart,
Rejoice in the Lamb, 1756-63

Traditional Common and Vernacular Names

Common names for plants, in addition to being easier to remember since they are in "plain English" rather than Latin, are often wonderfully descriptive, give us some interesting bit of information about the plant we might not otherwise have known, or spark our humorous imagination with such outrageous images as "Fairy-Elephant's-Feet". At the same time, common names can sometimes be confusing or misleading since the same name may be used for several completely different, unrelated plants or the same plant may be known by several totally different names. For example, of the twenty-four plants listed in *Hortus III* with "jasmine" in the common name, half are not of the genus *Jasminum,* and of the 206 "lilies" listed, only seventy-two are actually *Lilium.* On the other hand, the popular house plant usually known as a Ficus Tree is also known as Benjamin Tree, Weeping Ficus, Java Ficus, Laurel, Tropic Laurel, Weeping Laurel and Small-leafed Rubber Plant, and this just in the United States!

There is a wealth of common names for the oleander in other languages, too, as it is so well known in many different countries. In addition to being called "oleander" in many areas of the world, here are some vernacular names:

English:	Oleander, Rosebay, Rose Laurel
Dutch:	Laurierroos, Lauwerroos
German:	Lorbeerrose
French:	Fleur de Saint-Joseph, Laurelle, Le Laurier-rose
Spanish:	Adelfa, Alendro
Italian:	Alesswandrina, Allora d'India
Greek:	Agriodaphn, Nerion
Hebrew:	Ardaf, Harduf
Arabic:	Dafla
Sanskrit:	Asvamaraka
Hindi:	Kanel, Kaner
Tamil:	Agam, Alari
Bengali:	Karabi
Chinese:	Kap chukt'o (mingle-bamboo-peach blossom)
Japanese:	Kjotikto
Hawaiian:	Oleana
West Indies:	South Sea Rose
Puerto Rico:	Adelfa
Mexico:	Laurel, Laurel Rosa
Argentina:	Laurel Rosa

14

An interesting note from *Nerium News*, (Winter 1988): In Tuscany, Italy, the oleander is known as St. Joseph's Staff which was said to have burst into flower when the Angel announced that he was to marry the Virgin Mary.

In Nature's infinite book of secrecy
A little I can read, . . .
Shakespeare,
Athony and Cleopatra

Erroneous Common Names

While developing the Matrimandir Gardens in Auroville, India, we had the opportunity to collect seeds and plant material of many genera listed under the common name Oleander, some of which were truly oleanders and some not. Hundreds of years ago plant explorers and collectors did not have the benefit of ready reference to the work of others in the field, often identifying flora in the same country. The identification of taxa is easier today in our age of almost instantaneous data access and much of the confusion in synonymy of genera and species is being clarified as taxonomists access global databases. At the same time, we are experiencing an explosion of new plant cultivars and plants are being transported around the globe in ever greater numbers. Keeping up with the names of new cultivars as they are developed and different names for the same cultivars when they are introduced into different areas of the country, or into different countries, presents a challenge.

The following is a list of some of the plants commonly known as Oleanders:

Yellow Oleander, Be-Still-Tree	*Thevetia peruviana*
Climbing Oleander	*Strophanthus gratus*
Mexican Oleander	*Asclepias curassavica*
Wild Oleander, Water Oleander	*Decodon verticillatus*
Oleander Podocarp	*Podocarpus neriifolius*
Oleander Spurge	*Euphorbia neriifolia*

(Note the specific epithets for the last two in the list; "nerii", referring to *Nerium* and "folius" or "folium", referring to foliage, tell us that the plant has leaves similar to oleanders.)

There is a language, "little known",
Lovers claim it as their own,
Its symbols smile upon the land,
Wrought by Nature's wondrous hand;
And in their silent beauty speak,
Of life and joy, to those who seek
For Love Divine and sunny hours
In the language of the flowers.
J.W.H. (1913)

Commandant Barthélemy

3

*When worshippers offer flowers at the altar they
are returning to the gods things which they know,
or (if they are not visionaries) obscurely feel, to be
indigenous to heaven.*

Aldous Huxley

History and Geography

Distribution and Climatic Range

Tolerant of a wide range of soil and climatic conditions, oleanders are found naturally occurring near the Mediterranean Sea in northern Africa from Morocco through Algeria to Tunisia, in southern Europe along the Mediterranean coast from Gibraltar to Lebanon and Israel, and on islands in the Mediterranean such as Crete, Corfu, and Cyprus. Looking farther east, they are found again along the banks of the Tigris River from southeastern Turkey to the Iranian coast of the Persian Gulf, around the Gulf of Oman, inland in areas of Afghanistan and Pakistan, in India north of New Delhi, and in parts of western China and Japan.

A very adaptable species, oleanders are seen growing in open, full sun locations, in moist valleys along streams and seasonally dry watercourses, on sunny banks of mountain streams and in varying soil types. Usually indigenous in gravelly soils, they may also be found growing in sand and occasionally clay, tolerating both saline and alkaline conditions. Oleanders produce an extensive root system and are able to withstand long periods of drought as well as seasonal flooding. They grow in altitudes ranging from below sea level (in the Dead Sea area) to more than 8000 feet.

References to Oleander Cultivation in Ancient Times

Oleanders were found not only in ancient Greece but in Roman and Chinese gardens as well. "In China the cultivation of oleanders was a hobby of literary men who adorned their studies with cut oleander blooms," Pagen recounts in his book. "They especially appreciated the scent of oleander flowers and the elegant habit of the plant, and chose the oleander as an emblem of grace and beauty. To describe the oleander, the Chinese use three characters, successively meaning 'mingle,' 'bamboo,' and 'peach blossom.'"

Due to the preserving layers of volcanic ash from the eruption of Mount Vesuvius, we know that oleanders were also grown in the gardens of Pompeii. They were the plant most often painted on Pompeian murals (circa 79 A.D.) and were usually found represented in informal settings as background plants or mass plantings in the unique, traditional garden wall paintings whereby the Pompeians created the illusion that their gardens extended far into the countryside. In 1990 the Museum of Fine Arts in Houston, Texas, displayed an excavated wall preserved by the ash from Vesuvius. Similar garden wall paintings have been found in other places in Italy, including a garden room built by Augustus for his wife Livia, all of which include representations of oleanders.

Mrs. Isadore Dyer

Predating this, oleanders were widely planted in Rome during Cicero's reign, 106-43 B.C. Clarence Pleasants writes, "It was known by the ancient Greeks and was described by Pliny as having rose-like flowers and poisonous qualities." Theophrastus' reference to a plant he called *Oenothera* (circa 330 B.C.) is believed by authorities to be the oleander since he described it as a plant with a red, rose-like flower and leaves similar to those of the almond. He also mentioned its poisonous qualities. The name "oleander" is not specifically used in the Bible but is thought to be the "rose" mentioned in the Old Testament in references such as, ". . . it is like the rose that was planted on the river. . . ." The "willows of the brook," also believed to be the oleander, are mentioned in Leviticus 23:40.

The Hebrew holy text the Talmud mentions the oleander numerous times, especially in the Mishnah, the first part, compiled around 200 A.D. In Rashi's translation we read, "Because Moses sweetened the sour water with it, it was a miracle within a miracle, as the oleander was also bitter." In Israel the oleander has beautified the landscape for centuries, growing in all locales from desert to mountains, in the plains and along the coast. Its fragrant flowers are produced abundantly from spring into fall.

Pagen mentions E. Hyams as stating in *A History of Gardens and Gardening* that the oleander, the myrtle and the rose were the only flowering shrubs used by Arab gardeners of the Dar-al-Islam in the twelfth century. In *Tropica*, a contemporary horticultural encyclopedia by Alfred B. Graf, we find an extraordinary photograph of a red-flowered oleander growing in a moonscape desert setting. Graf's caption reads: "Nerium oleander, "Oleander" at home in arid mountains of Yemen, Southern Arabia."

Introduction of Cultivated Oleanders

The history of the oleander's introduction into Europe is chronicled in a number of texts. Clarence Pleasants writes that it was first introduced into western Europe by the early Phoenicians. F.J.J. Pagen relates that for many years the only form was the single, odourless, pink or red-flowered Mediterranean variety until a white-flowered form was found growing in the wild in Crete in 1547 and was introduced to Italy. From John Gerard's *The Herball, or Historie of Plants* (1597), we know that he was growing a red and white form in England in 1596. In the late 1500's and early 1600's oleanders gained popularity in England, primarily as conservatory plants to be moved out only during the summer months. John Parkinson writes in his *Paradisi in Sole, Paradisus Terrestris* in 1629, that a plant grown from seed brought to him in England from Spain by "Master Doctor John More" had a stem "as bigge at the bottome as a good mans thumbe" and that in 1640 it was "as bigge below as a reasonable man's wrist."

Mrs. Isadore Dyer

With the introduction in the late 1600's of single and double, pink-flowered plants from India and Ceylon (Sri Lanka), scented oleanders made their debut. These were quickly spread to all parts of Europe and various references are found to double-flowered forms in pink and red, and even variegated double flowers. Around this same time there is also mention of a variegated-leaved form of the Mediterranean single pink and a yellow-flowered form from India. In the late 1700's and early 1800's it became evident that crosses between the Indian and Mediterranean forms were being cultivated. The first known reference to an oleander cultivar is found in the French catalog "Buc'hoz" of 1799. It introduced *Nerium* 'Album', a vigorous and free-flowering variety. Pagen writes, "In 1840 Bosse mentions 36 cultivars; in 1849 he lists 58." During the late 1800's, Claude Sahut of Montpelier, France, worked extensively on oleanders and by 1898 had developed 170 cultivars.

Oleanders were first introduced to the United States in 1565 by Spaniards who founded St. Augustine in Florida. Early immigrants to America brought with them the double rose variety as well as white-flowered cultivars. These were handed down from generation to generation and treasured as a connection to their past. According to Elizabeth Head of the International Oleander Society, the oleander was especially popular among the Pennsylvania Dutch and could be found from the Shenandoah Valley to the valley of the Delaware and westward to Kansas. We know that early settlers carried plants with them to California where they found growing conditions in many parts of the state so ideal that oleanders are now used extensively for landscaping in low-maintenance situations such as highways and freeways. In fact, the longest flowering season on record is in the desert areas of California where oleanders can be found in bloom from May through December. Perhaps these

Mrs. Trueheart

same settlers, on route to California, were responsible for introducing the oleander to Nevada. Elizabeth Head reports that today oleanders are one of the most popular plants for landscaping and median plantings throughout Las Vegas. Many oleander varieties were brought from France and Morocco to Louisiana when it was still a French territory and they remain one of the most traditional of southern garden shrubs, gracing the formal but lush gardens of antebellum mansions as well as modern buildings in cities and towns along the gulf coast.

Our friend Jim Nicholas tells us that in the northeast, especially in the area of Connecticut where he lives, it is very common to see oleanders planted in large pots and tubs on decks or porches or even in front of shops where they bloom and grow magnificently. Observing that many residents are of Mediterranean origin, it seems likely that the plants are a reminder of their heritage and some have no doubt been propagated from cuttings brought to the United States from the old country by family members in years past.

The following vignette from Clarence Pleasants indicates the early presence of oleanders in the South Pacific:

> Historians trace the origin of Galveston's oleander to the South Sea Islands. Early in the British occupancy of the West Indies, the story goes, colonial governors sought to add the choicest fruits and the most beautiful flowers in the world to those already growing in profusion in the islands. Convinced that the soil and climate were unequaled, these gentlemen decided that the islands should be the beauty spot of which ancients had dreamed and masters of ships were charged to keep watch at every landing place for luscious fruits and beautiful flowers, and to bring back plants and seeds for transplanting. A captain of one of these vessels returning one day from the South Sea Islands brought with him a number of young oleanders. They were planted and the next spring the first oleander blossoms seen north of the equator added their delicate colors to the blooms of West Indian gardens. The South Sea Rose was the name then given it.

A friend of many years Genevieve Schutt who grew up near Sydney remembers oleanders being abundant in gardens in Australia, sometimes planted as individual shrubs and often as hedgerows. Her childhood memories include a large planting that surrounded a meadow in which she kept her horse. It was a single, rose-flowered variety common to the area with large clusters of intensely fragrant flowers which her horse would sniff, but seeming to understand the poisonous nature of the plant, never attempted to eat.

It is not clear at what period in history or by which route oleanders arrived in Central and South America but they have clearly flourished in many areas where the hot climate and arid conditions are ideal. We have seen them in full bloom in Mexico City, thriving in air pollution that had our eyes burning within minutes of landing at the airport. Around Cuernavaca, a lovely town in the mountains outside Mexico City where the days are hot, nights are cool and the air very dry, oleanders and bougainvilleas appeared to be favorite plants among the residents as there was hardly a garden without them, always in magnificent bloom. During a week spent diving on Ambergris Cay, a small island off the coast of Belize, we found oleanders thriving where few other plants would grow, not surprising since the island seemed made entirely of sand. On the Virgin Islands, too, oleanders are one of the mainstays in resort landscapes and private gardens alike, especially in the xerophytic conditions of some islands such as St. Maarten. On a recent trip which included Jamaica and Grand Cayman, we saw and photographed a wide variety of oleanders, from dwarf to tall, with many in full bloom in the middle of winter, some with ripe or ripening fruits, and others bearing both fruits and flower clusters at the same time.

Medicinal, Economic and Decorative Uses

The oleander in all its parts, roots, bark, leaves and flowers, has been utilized for medicinal purposes for centuries, especially in India and China. It has been used internally since early times as a cardiac stimulant having an effect similar to digitalis, as an antispasmodic, an abortifacient and as an antivenin. Its primary uses have been external applications as a salve (made from the leaves) and a paste (from the roots) to treat a variety of skin ailments including problems such as boils, rashes, lice, scorpion and snake bites, ringworm, scabies, headache, swellings, fevers, fun-

gal infections, leprosy and herpes. In folk medicine it was used to help build up strength, break the opium habit, stop people from drinking and treat malaria.

In Europe oleander leaves have been used for many years to expel fleas and baits employed as a rodenticide have been made from the bark and stems for more than a century. Dried oleander leaves are often laid amongst papers and books to keep insects away. In other countries the sap has been used as an arrow poison. In Crete the oleander is known to grow prolifically and attain such a size that its wood is used in construction.

Sue Hawley Oakes

Mrs. Robertson

Although oleander flowers tend to wilt fairly rapidly, with proper care they make excellent corsages and leis as well as lovely floral arrangements utilizing stems with full inflorescences.

Always cut flowers and branches in the early morning or, if this is not possible, late evening. Place branches in deep, warm (not hot) water in which a flower preservative has been added. These water-soluble products are effective in extending the life of cut flowers and we recommend them highly. Be sure to remove all leaves below the water line to prevent them from rotting. Do not use water treated with a water softener. Place cut flowers in the refrigerator or a cool place until ready for display.

As with plumerias, there are many ways to fashion leis and garlands using a needle and thread. Indulge your creativity by mixing flowers of different colors, or even different kinds of flowers with the oleander blossoms. As a general rule, single flowers are more satisfactory since they tend to remain fresh longer than double flowered forms. One should experiment with a number of cultivars to determine keeping quality. Good flower substance (thickness of the petals) is often an indicator of longer lasting blossoms.

Cultural and Religious Symbolism

People have used oleander flowers for making leis and garlands for centuries. This was a tradition during Roman times and is equally so today in countries as far apart as India and the United States. In the 1800's, when oleanders were a status symbol, corsages and clusters of the blossoms were popular among ladies who used them to adorn their dresses.

According to Marie C. Neal in her book, *In Gardens of Hawaii*, "The Moors attribute great magical virtue to 'the sultan of the oleander', which is a stalk with four pairs of leaves clustered around it."

Clarence Pleasants relates, in his inimitable style, this poignant story about the oleander in his pamphlet on the genus, written March 17, 1967:

> There is a touching legend about the oleander and a pair of young lovers. Leander went to sea and his sweetheart would walk along the seashore watching for the return of his ship. After a dreadful storm she learned his ship had been wrecked and all on board lost.
>
> On moonlight nights the sad young girl would walk among the flowering shrubs in her father's garden crying, "Oh! Leander," over and over again. Feeling life unbearable without Leander she picked a spray of exotic shrubs and walked into the sea. On finding her with a spray of flowers clutched in her hand people began calling the shrub Oh! Leander.
>
> With its exotic loveliness and heady fragrance, sometimes intoxicatingly sweet, sometimes almost bitter, this old world flower has a cruel duplicity, for in its stalk, leaves and even in the delicate blossoms is contained a poison which in large quantities can be fatal to men and animals.
>
> To be lovely, to be admired is its sole purpose and in this it contributes indescribably to man's pleasure.

In an excellent article in *Nerium News* (Fall '90), entitled "Legends of the Oleander," Dr. Edith Box offers this version of the same story:

Henry Rosenberg

In keeping with the Mediterranean origin of the oleander, one legend has it that oleander in Greek mythology means romance and charm. A beautiful Greek maiden was wooed by Leander who swam the Hellespont every night to see his beloved. One night he was drowned in a tempest. Wild waves dashed his body against sharp rocks and left him lifeless on the white sands. Here his lover found him as she walked the shores calling "Oh Leander, Oh Leander." A beautiful flower was clutched in his hand. She removed it and kept it as a symbol of their love. Magically it continued to grow and from this symbol of everlasting love came the beautiful and abundant oleander.

Pink Beauty

Dr. Box continues with several more stories that we are pleased to include:

In 1915, Charles M. Skinner published another legend concerning St. Joseph and the oleander in a local paper. It seems that a poor but lovely Spanish girl lay ill of a fever. Her mother tried all that her meager resources permitted to cure her daughter, but to no avail. Exhausted by her desperate efforts, the mother fell to her knees to pray to St. Joseph to spare her child. When she arose, the room was filled with a rosy glow from a figure bent over the girl. A strange man of noble demeanor placed on her breast a branch consisting of pink unfading flowers from paradise. As the rosy light faded, the mother rose to thank the noble stranger but he had disappeared and she was alone with her daughter who was sleeping soundly for the first time since she became ill. The mother again knelt to give thanks and since that day the oleander has been known as the flower of St. Joseph.

. . . .

Another legend involving the buccaneer, Jean Lafitte, in the establishment of oleanders on Galveston Island is less romantic. In the pursuit of his pirate's craft, Lafitte had attacked a Norwegian schooner and killed all the passengers except for one man who was clinging to a beautiful flowering plant. His name was Ole Andersen; Lafitte saved him and made him his gardener calling him Ole Ander. He later honored him by calling his flower by the same name.

. . . .

Two traditional uses of the oleander come from different parts of the world. In India, Hindu mourners place oleanders about the bodies of dead relatives using the blooms as funeral flowers. In Germany, on the other side of the globe, women have a tradition of placing potted oleanders outside the kitchen window for good luck.

Oleander flowers are used in worship in many parts of the world. The Hindus offer the flowers to Shiva, the god of destruction and regeneration. According to Pagen, "In India, Italy and Greece the oleander is associated with and used at funerals. . . ;" and, "It is the floral emblem of Saint Joseph, and is popular for making temporary shelters used in the observance of the Jewish Feast of the Tabernacles." In a letter published in the Spring 1989 issue of *Nerium News*, Christine Pape in Evanston, Illinois, says that in a neighborhood of Italians, Poles and Germans in Chicago there is a traditional ceremony of gathering together each fall to carry large, potted oleanders indoors to the basement for the winter, and again

Agnes Campbell

26

in the spring to carry them back outdoors for the summer. Jim Nicholas relates a similar story about a friend of his who was born in Hungary. Her childhood memories include a big oleander growing in a tub that her father would carry to the basement each fall where it lived through the winter until brought outdoors again each spring.

Rabindranath Tagore, the beloved Indian poet, writer (winner of the Nobel Prize for Literature in 1913), musician, philosopher, mystic and founder of Shantiniketan, a highly acclaimed school of music in northern India, wrote a play called *Red Oleanders*, first published in 1925. In it the heroine is Nandini, a young girl who refuses

Algiers™

to bow to conventional social mores in a village where men are enslaved to mine gold. Through her courage and inspiration, which Tagore symbolizes with the red oleander, the men awaken to their plight and also to the idea of freedom.

In Xanadu did Kubla Khan
A stately pleasure-dome decree
Where Alph, the sacred river, ran
Through caverns measureless to man
Down to a sunless sea.
So twice five miles of fertile ground
With walls and towers were girdled round:
And there were gardens bright with sinuous rills,
Where blossomed many an incense-bearing tree;
And here were forests ancient as the hills,
Enfolding sunny spots of greenery.

Samuel Taylor Coleridge,
Kubla Khan

27

Martha Hanna Hensley

4

It's funny how you could see something all of your life,
And then one day really see it.

Dr. Barry Comeaux

Description

Plant Types and Structure

The most natural form of an oleander is globular, or upright and rounded, whether it be a dwarf or tall-growing plant. The foliage is normally full and dense, skirting the ground. With many new oleander hybrids appearing each year, a criteria for grouping by size as well as color seemed useful to determine varieties for foreground, middle, background or specimen use. It would also assist gardeners in colder climates by giving them a more accurate guideline for choosing plants to be grown in containers. After viewing thousands of oleanders around the world, we evolved a classification system of five categories of size — only to find that Clarence Pleasants had already defined an almost identical system! We have combined the two systems for the following classification of mature plants:

Miniature	1 to 2 feet maximum height
Petite	2 feet to 5 feet maximum height
Dwarf	5 feet to a maximum of 8 feet
Intermediate	8 feet to a maximum of 12 feet
Tall or Large	12 feet to over 20 feet

Growth Habits and Life Span

Oleanders are moderate to fast-growing shrubs that are relatively long-lived. Plants tend to send up numerous stems from the base in the larger forms, but growth is much more controlled in dwarf and petite varieties. There are named cultivars to suit every landscape situation from slow-growing, dwarf, rounded, tightly compact forms to tall shrubs that can be trained in a multitude of shapes such as standards (also known as single trunk or patio trees), multi-trunk or multi-stem specimens, and a variety of other forms that are fully covered under Pruning in the chapter on "Culture."

Stems and Branches

Stems are basically upright with few branches, initially a medium green chang-ing to grayish tan on maturing and with a clear (not milky), sticky sap. Branches are flexible and bear mostly terminal inflorescences subtended by three equal branchlets (slightly triangular at the apex), giving the appearance of a candelabra shape.

Ted Turner, Sr., owner of Turner's Gardenland in Corpus Christi, makes an interesting observation regarding growth characteristics of oleanders in the Cor-pus Christi area and southern Texas, in contrast with southern California and other parts of the world. According to Ted warm summer nights tend to elongate the stems between leaf nodes and the leaves themselves are larger. In a compari-son of like varieties, the overall tendency is for plants in Corpus Christi to be slightly larger and more upright growing. This is a concept often overlooked in general horticultural literature.

Leaves

Oleander leaves are simple, narrow, lanceolate, thick and leathery, usually smooth and glabrous above, rarely puberulous beneath, and cuneate at the base with short petioles and a prominent mid-rib with faint lateral veins. Margins are entire and slightly revolute. Leaf color ranges from rich, deep green to gray-green above and lighter beneath. Leaves vary in length from 2 to 3 inches to more than 12 inches. Although the leaves are somewhat stiff and leathery, their shape and arrangement on the stems gives plants a fine to medium-textured appearance.

Turner's Sissy King™

30

Tangier™

Inflorescence

Inflorescences are mostly terminal cymes, variable in size, bearing loose or dense clusters of flowers with sepal-like bracts beneath. Blooming occurs on the new growth with one cyme often producing as many as fifty to seventy-five flowers.

Flower Shapes, Sizes and Colors

Oleanders offer flowers in a wide range of colors and hues, in fact, almost every color of the rainbow is represented with the exception of the blue-violet spectrum. We have had reports of lavender flowers in Europe but will have to visit the oleander collections there and see for ourselves! There are great variations in the many shades of red and pink, especially when the corona is of a different color.

The satiny blooms are generally large and showy and may be fragrant, often highly so, or totally without scent. When flowers are pale to white, sepals are light green. In deeper colored flowers, sepals are usually dark red to burgundy. The corolla consists of five overlapping lobes that are twisted to the right in the bud stage. The stamens are attached at the apex of the corolla tube but are not exserted. The anthers are adhered to the stigma.

Flower Shapes

There are three basic flower shapes with variations within each shape.

1. Single — Five fused petals (corolla lobes) with a central corona, a corolla tube, and a calyx with five sepals. The petals may be pinwheel, windmill, star-shaped, or cup-shaped, with the arrangement of the corolla lobes varying from narrow and widely separated to broad and overlapping.

2. Superimposed corollas — The technical term for semi-double or hose-in-hose flowers. In this shape one corolla is superimposed inside the other with the corolla lobes of the two whorls alternating.

3. Double flowers — This is a complicated and irregular form and we have followed F.J.J. Pagen's definition:

> The term "double" in describing a flower form is ambiguous, but in this classification it is defined by the following characters (not all characters may be present in the same flower):
>
> > - More than ten petals or petaloid parts.
> > - Irregularity.
> > - Some petals free, some fused.
> > - Petals with claws.
> > - Petaloid anthers.
> > - Broad, swollen buds.
>
> The double oleander flower has an irregular structure. Although the flower basically is composed of two, three or even four whorls of petals in addition to the calyx, the system is usually distorted and often hard to unravel.

We have found considerable variation in the double flowers of many plants; blooms on the same plant are often single, others hose-in-hose, and yet others double but with irregular differences even among the double forms.

Flower Sizes

Corollas are variable with average sizes ranging from 1 inch to 3 inches or more.

Flower Colors

Colors range from white to pale pink, many shades of medium to deep pink, bright red or cerise to deep red, pale yellow, medium yellow and many salmon shades, often with orange coronas. A strongly colored corona tends to significantly influence the apparent color of a single flower, especially if the corolla lobes are somewhat cupped. The coronas often have an apical fringe and frequently exhibit vertical lines of a different shade. Coronas that are strikingly colored or striped impart different hues not only to the individual flower but to the entire inflorescence.

Fragrances

As we mentioned in the "Introduction," virtually all the oleanders we grew in India were highly fragrant, often delightfully so. There are a number of fragrant varieties grown in the United States but they are few relative to the number of cultivars. Oleander fragrances are nearly as difficult to identify as those of plumerias. Many of the varieties in India, and some in the United States, have a jasmine-like fragrance, others a sweet-spicy or musky scent. Fragrance is very subjective but among the people we interviewed the scent most often described for oleanders in the United States is that of vanilla with certain cultivars imparting a magnolia or bitter almond aroma. Ted Turner, Sr. stated that he has found all of them to be most fragrant in the morning, becoming less so as the day proceeds.

Turner's Elaine Turner™

Follicles and Seeds

Fruits are more readily produced on the fragrant varieties and free-blooming cultivars than on other varieties. The seed pods (technically bicarpellate follicles) look like bean or pea pods in the shape of two narrow downward curving horns and are subtended by the persisting calyx. (See photo pg. 80) They are variable in length from just over 2 inches to approximately 7 inches and are usually about ½ inch across.

The follicles vary from dull or bright green in the immature stage to medium brown or reddish brown at maturity. One often finds them in the process of maturing with the tip turning brown and the end still green where attached. When ripe and beginning to dry, the follicles split longitudinally to reveal numerous light to dark tan seeds (usually at least 50, often up to 200) covered with very fine hairs and a coma (tuft of hairs) at the apex allowing for wind dispersal.

Mrs. Lucille Hutchings

... How love burns through the putting in the seed
On through the watching for that early birth
When, just as the soil tarnishes with weed,
The sturdy seedling with arched body comes
Shouldering its way and shedding the earth crumbs.
Robert Frost

5

*There is no monotony in flowers, they are ever unfolding
new charms, developing new forms and revealing new
features of interest and beauty to those who love them.*

Joan Wright

Selected Oleander Cultivars

It would be almost an impossible task to attempt to photograph and describe in detail all the oleander cultivars in existence today, many of which are commercially unavailable. It would also lead to a massive, unwieldy tome that no one would read! While only about fifty varieties were offered in southern nurseries in the United States during the 1940's, our most recent estimate of varieties in cultivation is between 400 and 500. In his book *Oleanders: Nerium L. and the oleander cultivars*, F.J.J. Pagen includes a section entitled "Tentative Checklist of Oleander Cultivars" in which he reviews plant catalogs from Belgium, France, India, Italy, Japan, Portugal, Spain and the United States, spanning a period of more than 100 years. His "Checklist" of 599 different cultivar names, in which he notes numerous instances of misspellings and synonyms, ultimately represents 401 distinct cultivars for which he provides information on flower type, color, scent, origin and date of origin, as well as the first published reference and its date. This alone occupies 59 pages, more than half his book! Pagen mentions that approximately 175 varieties are available in the nursery trade today (primarily in Europe).

Extensive propagation and hybridizing work are being carried out in Italy. Michele Caponigro, a student in Turin who is writing a thesis on the oleander, has recently corresponded with the International Oleander Society (*Nerium News*, Spring 1996). From his letter we learn that southern Italy exports around 50,000 potted oleanders a year to northern Europe where they are becoming increasingly popular as house plants. Varieties such as 'Papa Gambetta,' 'Maria Gambetta' and 'Pietra Ligure' are some of the new cultivars recorded in the mid 1970's and 1980's.

Oleanders are also widely grown in Israel today and many dwarf as well as hardy forms have been introduced. *The Nerium Oleander in Israel*, by D. Zafrir, describes and pictures twenty named cultivars. (There is also a factory that extracts cardiac glycosides from oleanders.)

Our handbook is intended to inspire gardeners to cultivate oleanders in all areas of the world. In this first edition we have chosen to highlight some of the latest hybrids, including a few that will soon be released, and concentrate our focus on varieties grown in the United States that are, to a greater or lesser extent, available to

35

the gardening public. It has been our experience that few things are more exasperating to plant collectors and gardeners than to read about a delightful and desirable group of plants and not be able to find a source for them!

One note that bears repeating is the need for documentation. From personal plant collections to public botanical collections, documentation and labeling are of paramount importance. Clarence Pleasants even made the plea to members of plant societies who discover something unique in a neighborhood garden, to make contact with the owner, impress them with the fact that they have a unique plant, invite them to a meeting of your society and ask them to label the plant so that rare specimens may not be so easily lost as they often tend to be in America's rush to build more freeways and sub-divisions. In many states the Department of Transport will inform native plant societies of new road construction. These societies then organize "Plant Rescues" to collect and preserve any rare and endangered species that otherwise might be destroyed.

In many cases a plant name is unknown and the first step in documentation is identification. Until "gene mapping" becomes more readily available and less costly,

we must accomplish this by physically comparing plant and flower characteristics to other plants or herbarium specimens that are already positively identified. In doing this it is important to be aware that even named cultivars change their flower colors according to the season; some deepen in hue as the season progresses and others become lighter. Colors are also affected by the intensity of the sun or the lack of it; many plants have a deeper color on cloudy days or when grown in filtered light or part shade. The time of day can also be a determining factor. With the advent of cooler weather the colors of many cultivars deepen. This is true with many other flowers

Hardy Red

as well, especially hibiscus. Some authorities suggest that the type of fertilizer applied may also change the color of the flower. Clarence Pleasants once remarked that the colors of many oleanders grown in the city of El Paso appeared much more intense. He believed this to be caused by something in the soil as well as the dry air and intense sun.

Listed below are the basic colors ascribed to each cultivar as it is typically seen. Since there is a plethora of synonyms and names of no horticultural standing, especially regarding cultivars in the United States and Galveston in particular, we are truly indebted to Bob Newding for his systematic observations over the past two years. We have augmented our own studies and observations with his excellent descriptions of many of the oleanders found in Galveston. Our thanks is also due to Clarence Pleasants who helped us with many descriptions.

Hardy Red

Oleander Cultivars in the United States

Red Flowers

Single Red Flowers

'Algiers'™ 'Monal' — A Monrovia introduction from northern Africa (1978). Single, deep red, almost fluorescent flowers, 2 to 3 inches across, are borne on dwarf, free-blooming plants from early spring to first frost, providing continuous color.

Although not as prolific in bloom as the petites, the variety is nonetheless stunning due to the extraordinarily distinctive color. An excellent choice for containers and low hedges. Growth is moderate to fast in the landscape with plants attaining a height and spread of 8 to 10 feet at maturity. A half-hardy variety. (See photo, pg. 27)

'**Calypso**' — Eyecatching, iridescent, cherry red blooms. According to Monrovia Nursery in California the plants have been observed to be hardier than 'Pink Beauty' or 'Hardy Red.' It blooms all summer and is Bob Newding's choice for the best red for Galveston. We have seen a striking specimen grow-

Photo courtesy Monrovia Nursery Company

Ruby Lace™

ing at the water's edge on Galveston Bay adjoining Newding's house, demonstrating that the plants are very salt and wind tolerant as well.

'**Hardy Red**' — One of the hardiest oleanders known, withstanding numerous freezes without damage. In the most severe freezes it will be killed to the ground, but with the advent of the first warm weather new, lush growth will rapidly regenerate. Newding observes that flowers have a prominent, white, apical fringe on the anther. (See photos, pgs. 36, 37)

'**Little Red**' — (Plant patent #4836) Patented by Aldridge Nursery, Von Ormy, Texas, (1982); the parentage was single 'Hardy Red' x 'Scarlet Beauty.' The seed from the cross was collected in 1958 and planted. Of more than 200 seedlings transplanted into the open field in 1959, all but three were inferior to the original parent plants. One of the three appeared to be more dwarfed and cold hardy and was selected for further study. The remaining seedlings were destroyed during the winter of 1966. The flowers of the selected plant are the same deep color as 'Scarlet Beauty' but are smaller than either parent and the plant grows one-half or less the size of 'Hardy Red'. Although of excellent color, this variety tends to be a bit temperamental and along coastal areas seems to be more susceptible to diseases, windburn and salt damage to the leaves.

'Jannoch' — Listed in Pagen's book as first published in the 1952 catalog from Monrovia Nursery. The flowers are red with a red corona that is highly fringed. The corolla lobes are widely separated. Clarence Pleasants described it as a "big, wide flower" and added that it was a good bloomer, covering the bush with a blanket effect rather than with clumps and clusters. He also noted that plants tended to be rounded, somewhat compact and hardy.

'Marrakesh'™ 'Moned' — A special selection from seedlings grown at Monrovia Nursery and introduced in 1994. The shape is a mounded form and, according to Monrovia, grows only to a height of 5 to 7 feet. It blooms throughout much of the year in California and produces masses of exotic, rich, warm red blooms. Newding says there are very few dwarf reds and this is one of them. He considers it far superior to 'Little Red' but since it is so new more study is needed. (See photo, pg. 40)

'Mrs. Robertson' — A very large, fast-growing variety that flowers in spring. The blossoms are large and pinwheel shaped, deep cerise with a corona of the same shade. A half-hardy plant with very fragrant flowers. (See photo, pg. 23)

'Professor Parlatorre' — In some ways similar to 'Mrs. Robertson'. According to Bob Newding, it is an "uncommon, large, cerise, pinwheel flower with an unusual lavender cast." A large, half-hardy plant that flowers heavily in the spring.

'Ruby Lace'™ 'Monvis' — A 1986 Monrovia introduction. A very unusual cultivar with large, 3-inch, pinwheel flowers of intense ruby red. The corolla lobes have scalloped edges with a fringed lip that create the "wavy" appearance of the flower. Intermediate in size, growing to 12 feet in height and as wide with dense, compact foliage. (See photo, pg. 38)

'Scarlet Beauty' — Another favorite of Clarence Pleasants who described it as "a light, showy red, it catches your eye, a different form of red." Newding notes its "deep cherry red, single flowers, at times with blue-black margins with a velvet sheen. A beautiful spring bloomer that reblooms sporadically throughout the summer." A half-hardy plant with a large, upright form. (See photo, pg. v.)

'Sugarland'™ — (See 'Hardy Red' for description.) An introduction from Hines Nursery discovered by their horticulturist Bill Barr. The variety was found amongst a group of oleanders (all red) in a traffic median in Sugarland, Texas. Bill observed the group of plants for two winters. The plant he selected for hardiness was untouched during freezes that leveled all the other plants in the median to the ground. Subsequently, during the freeze of 1983 (single digit measurements for days) while all other oleanders were killed to the ground, 'Sugarland', again, did not even have leaf burn. Hines Nursery had many plants of the variety known as 'Hardy Red' which froze, but herein lies a problem that can only be solved eventually by gene mapping. According to Bill Barr, the plants that froze and were labeled 'Hardy Red' may have been from an unreliable source and might not have been 'Hardy Red' at all. Some members of the International Oleander Society believe 'Sugarland' to be the true 'Hardy Red' as growth characteristics and flower color are identical.

Double Red Flowers

'**Commandant Barthélemy'** — A freeze tolerant old cultivar with large, highly fragrant flowers that are light to medium red with some streaks of white. Clarence Pleasants found it on the outer banks of North Carolina around 1956 where it grew lushly in an area of milder winters. A free-flowering plant originated by Sahut in Montpelier, France, and first published in the catalog, *Sahut*, in 1898. (See photo, pg. 16)

'**General Pershing'** — Even among oleanders this is a unique plant. It is grown not only for its flowers but for its long, stiletto-type leaves which attain a length of 12 inches in full sun and even longer in part shade. According to Pleasants the leaves are often twice as big, though not quite twice the width, of most oleanders and distinguish the plant even when not in bloom. It develops a mounding form, tending to weep and spill over rather than grow upright. The flowers are large and a very attractive dark red. Pleasants said it flowers heavily, others note that flowers are borne sparsely. Very hardy. (See photo, pg. 73)

'**Mrs. Kempner'** — According to Bob Newding, this plant was thought to be lost after the 1983 freeze. Since then two plants have been found that match the description. Newding describes it as ". . . a gorgeous plant with very rare, deep rose-red, triple, carnation-like flowers." Half-hardy.

Photo courtesy Monrovia Nursery Company

Marrakesh™

Pink Flowers

Pink seems to be one of the most popular colors in the United States and there are more pink oleanders than any other color.

Single, Light Pink Flowers

'**Apple Blossom**' — In a letter to Mrs. Corinne E. Kirchem of Galveston, Texas, May 29, 1978, Mr. R.C. Aldridge, Jr., President of Aldridge Nursery, describes the genesis of his cultivar as follows:

> Our nursery first grew the Sealy Pink variety in 1936. Our stock was secured from a nurseryman in Corpus Christi, since deceased.
>
> The Sealy Pink was frozen back each winter in our area and northward so I attempted to create a hardier plant with the same blossom color. In 1939 I hand pollinated a Cardinal variety (Single Hardy Red) with Sealy Pink. Seedlings from this cross were planted in the field in early 1942, about two thousand in all. In 1946 it was apparent that only one of the complete lot was worthy of selection and propagation. When we first offered it to the trade we called it Apple Blossom. It proved to be much hardier than Sealy Pink and has been widely distributed from California to Florida.

'Apple Blossom' is an especially beautiful cultivar with attractive pink flowers in large clusters on a large, rounded, spreading shrub. Kewpie Gaido gave us a

Ella Sealy Newell

41

plant some years ago and said it was one of her favorite oleanders. Pleasants mentioned that it grows well in a container and cited an unusual quality:

> It can be cut back. When I was at the Flagship for Mardi Gras one time, they wanted me to cut back the 'Apple Blossom' which was overhanging the area where the announcers were going to sit. So, I trimmed it neatly back and then, when I came to work somebody had trimmed it very severely because the Channel 13 crew was going to sit behind it. I said, "That's the end of the blooms." Well, around July when all the new growth came out it had more flowers than usual. It bloomed just the same but even prettier than before. It's a very attractive plant and would make an excellent tub plant, a good driveway plant or a one-and-only [specimen]. If you had to choose one pink, this would be good. (See photo, front cover)

'Barbara Bush' — Named after the First Lady, this is an especially lovely cultivar that blooms all summer with light blush-pink flowers. The plant is intermediate in size and was a favorite of Clarence Pleasants who commented, "She [the First Lady] didn't give us much publicity. I knew she wouldn't. I imagine they named some aluminum wear after her too. You can't promote all that!" Pleasants observed that the plant remained in bloom throughout the year with a scattering of blooms even in December and January. "It just doesn't want to give up," he wrote. "On the first warm day in spring they'll be some of the first blossoms you see. It's a bushy oleander and it's covered with blooms spring and summer." (See photo, pg. 12)

Frances Moody Newman

'Lady Kate' — Bob Newding describes it as having ". . . large, single, pinwheel flowers that are light blush-pink with ruffled edges. The ruffled edges are important and the original description says the lightest of pinks. It's a prolific spring bloomer on a large, half-hardy plant . . . it doesn't have a lengthy blooming time but the entire plant is covered with flowers and it's very attractive."

'Mrs. Trueheart' — A magnificent free-blooming cultivar with large, fragrant, single pink flowers. Pleasants rated it among the top performers and described it as having ". . . big flowers and the stripes in the center are very prominent; it's very fragrant — that turns most people on, me too, and it's bushy, more rounded." Kewpie Gaido also lists it high among her favorites for the flower size, fragrance and lovely shade of pink. (See photo, pg. 20)

'Petite Pink' — An excellent free-blooming plant for container culture as it flowers prolifically with masses of soft shell-pink blooms and will remain compact if its root system is restricted. In the ground in Galveston it can achieve a height of 8 to 10 feet but with pruning is easily kept in the 4 to 6 foot range. The cultivar was introduced by the Los Angeles State and County Arboretum. (See photo, pg. 93)

Single, Medium Pink Flowers

'East End Pink' — This variety is often sold by nurseries as 'Sealy Pink' (the so-called generic 'Sealy Pink'), but is also sometimes sold as 'Pink Beauty' (an example of the license nurseries often use in naming their plants). 'East End Pink' is one of the most floriferous spring bloomers, is very hardy, and is commonly found on the east end of Galveston Island. The corolla lobes are reflexed and the pink flowers have a bluish cast to them. (See photo, pg. 144)

'Ella Sealy Newell' — An uncommon plant bearing medium pink flowers with yellow-orange at the base and yellow coronas striped pink. The single flowers are arranged in clusters resembling hydrangeas. Blooms in spring and often throughout the summer. (See photo, pg. 41)

'Frances Moody Newman' — A very attractive, half-hardy variety that flowers all summer and is widely planted around Galveston in malls, shopping centers and along highways. The single flowers are a rich deep pink with an iridescent cast. Newding notes that the corona is creamy white with pink stripes. It is named after Mrs. Newman who served on the Board of Directors of the Moody Foundation in Galveston. (See photo, pg. 42)

'George Sealy' — A beautiful plant but rare because it is tender. This was another favorite of Clarence Pleasants for its fragrant, deep pink flowers with striped, deep reddish-pink coronas. It flowers in very attractive clusters and is good as a pot plant since it is medium-sized and very floriferous even in small containers. (See photo, pg. 125)

'Hardy Pink' — Bob Newding believes this to be a generic name for 'Pink Beauty' and therefore a name of no botanical standing.

'Henry Rosenberg' — An unusual variety bearing medium pink, star-shaped flowers with thin, rose-pink stripes in the throat. Plants are large, half-hardy and bloom primarily in the spring. (There is a good specimen in a planter at the south entrance to the Rosenberg Library in Galveston.) (See photo, pg. 24)

'Kewpie' — The story of the discovery of the oleander named 'Kewpie' was related to us by Clarence Pleasants and is illustrative of how many superior plants have resulted from chance seedlings. Ethyl May Koehler, Kewpie Gaido's close friend, found an unusual plant whose flowers reminded her of Kewpie and after receiving permission from the owner, Clarence helped dig it up. The flowers were pink with some blossoms strongly variegated; one branch consistently produced flowers that were all variegated. This sport was rooted and named in honor of Mrs. Kewpie Gaido. It is a large pinwheel-shaped flower, variegated pink and white. Plants are free-blooming and half-hardy. (See photo, pg. vi.)

'Martha Hanna Hensley' — One of the most beautiful cultivars with delicate pink and white variegated flowers borne in large tight clusters and emitting a light fragrance. Plants are compact and bloom at a very early age. (See photo, pg. 28)

'Mrs. Masterson' — Delicate, pink, bell-shaped flowers with creamy white throats marked with light rose-pink stripes. The calyx is yellow-green without prominent sepals. Plants are large and bloom early in the season.

'Pink Beauty' — A spring blooming plant that has large windmill-shaped flowers and extraordinary, swollen, swirled buds shaped like parasols just beginning to open. Pleasants liked it for a number of reasons, among these the fact that it makes an excellent standard, adapts very well to diverse cultural conditions and has large

Mrs. Eugenia Fowler

44

flowers that simply cover the plant at its peak of bloom. Introduced by Monrovia Nursery in California in their 1952 catalog, it makes a superb windbreak and is especially suited for large hedges. Often sold under the name 'Hardy Pink'. (See photo, pg. 25)

'Pleasants Postoffice Pink' — Named by the International Oleander Society to honor Clarence Pleasants who discovered the plant growing on Postoffice Street in Galveston, Texas. The large, single flowers are medium to dark pink and are borne from early spring into summer on large plants. Bob Newding tells us the only plants in cultivation are found on the northwest side of Open Gates, the former Sealy mansion, in Galveston. (See photo, pg. 133)

'Sealy Pink' — 'Sealy Pink' is a generic term used by nurseries to refer to any single pink oleander and is a name of no botanical standing according to Newding. He further notes that the plant usually described as 'Sealy Pink' is actually 'George Sealy'.

Mrs. Eugenia Fowler

'Tangier'™ **'Monta'** — A Monrovia introduction from northern Africa in 1978, this is beautiful free-blooming cultivar with soft pink, single flowers in spring. As summer progresses the flowers change to a medium pink. Moderate to fast growing; becomes 8 to 10 feet tall in the landscape. (See photo, pg. 31)

'Turner's Flirt'™ — This dwarf to medium-sized Turner hybrid blooms eight to nine months of the year. Upon opening flowers are an iridescent hot pink with red stripes and yellow throats and later fade to light pink, giving the clusters a lovely shaded quality. One of the most unique characteristics is the scent which is reminiscent of double-bubble gum! (See photo, pg. 54)

'Turner's Tickled Pink'™ — (Patent pending 5/1/90) A Turner hybrid that offers many unique qualities. Turner describes it as having single, light peach-pink, candy-striped flowers with red-yellow coronas. The corolla lobes change to pinkish-lavender or orchid at the outer edge giving the flower a pinwheel effect. Medium-sized clusters are borne throughout the plant. This is another good choice for containers with its compact, upright growth and free-blooming habit. In the ground it can reach 4 to 6 feet but can be easily pruned to less than 4 feet, (although some plants in Galveston have reached over 12 feet). Hybridized by Ted L. Turner, Sr. and named by his daughter Shari. (See photo, pg. 69)

Double and Semi-Double (Hose-in-Hose) Pink Flowers

'John Samuels' — A very rare variety with fully double, rich deep pink flowers with slightly ruffled edges. Named in honor of the Samuels family in Galveston, there is only one plant known to exist on the island. (See photo, pg. 96)

'Mrs. Burton' — Gorgeous, fragrant, rose-pink flowers are both double and semi-double and somewhat ruffled. A rare plant, very desirable and very beautiful, blooming all summer. Pleasants recommended it highly, remarking that the pink is striking against the dark green leaves, and suggested that it would be a good plant to grow in more northerly areas because it offers the added bonus of being very fragrant. (See photo, pg. xiv)

'Mrs. Eugenia Fowler' — A late spring-flowering cultivar that blooms prolifically with flowers literally covering the plant. Its branches will actually weep from the weight of the blossoms. This is a large, hardy variety that Newding remembers being ". . . planted around many Galveston public schools prior to 1950. Those plantings survive at the site of the former Crockett School, now a Galveston College parking lot at 39th and Ave. Q." (See photos, pgs. 44, 45)

'Mrs. Isadore Dyer' — A fully double pink that came to Galveston from Jamaica in 1841, this variety blooms almost all summer, is hardy and has very fragrant flowers. In spring, blossoms are light pink with a single, white stripe through all the petals. As hot weather increases the flowers turn a much deeper pink. Pleasants noted that it is often sold in nurseries as 'Double Hardy Pink'. (See photos, pgs. 18, 19)

'Mrs. Runge' — A most unique cultivar with its large, fragrant, double, deep rose-pink flowers that stand out against the strikingly variegated, yellow and green leaves. It must be grown under good conditions or will be temperamental since the leaf area has so little chlorophyll due to the variegation. (See photo, pg. 66)

'Mrs. Swanson' — A double, light pink cultivar, blooming from spring through summer. Originally from California, it has evidently traveled widely as Clarence Pleasants brought it from Virginia to Galveston, Texas. He considered it to be even hardier than 'Mrs. Isadore Dyer'. Growth is upright rather than spreading.

Salmon Flowers
(Includes salmon-pink, apricot and peach shades)

Single Salmon Flowers

'Agnes Campbell' — A vigorous plant that flowers in the spring with attractive, large, salmon-pink to creamy salmon, pinwheel-shaped flowers with yellow coronas. Bob Newding calls it a half-hardy plant with a vigorous growth habit and says, "I mean vigorous! It grows faster than any oleander I have in my yard." For all its vigor, plants grow compactly with dense foliage. (See photo, pg. 26)

'Franklin D. Roosevelt' — Named in honor of President Roosevelt on his visit to Galveston in 1938, this very hardy, intermediate-sized plant thrives in Galveston, growing in several of the fishing camps along the waterfront and in the esplanade on I-45 entering the city. Flowers are a unique salmon-orange that turn almost coppery orange as the season progresses. Newding says of it, ". . . people comment and say, 'What is that beautiful oleander?' Grown in mass plantings it is an incredible splash of color and one that will bloom from spring to mid-summer — which is fairly long, even though it's not a free-bloomer." Pleasants considered it a good container plant and said it blooms more than 'Hawaii'. (See photo, pg. 114)

'Hawaii' — A favorite of John Kriegel, Director of Gardens at Moody Gardens in Galveston, it is one of the loveliest of all oleanders but one of the most tender. (Newding lists 'Kathryn Childers', 'Hawaii', and 'Casablanca' as the most tender plants in his yard and the only plants that showed any real damage in 28 to 30 degree temperatures.) A very attractive, almost free-blooming plant with large, single, salmon-pink flowers with somewhat pinwheel-shaped corolla lobes, widely separated and squared at the tip. The yellow-gold throat is marked with faint pink stripes. For best performance Newding recommends planting in a protected place with southern exposure. An excellent plant for container culture. We observed it in striking bloom in Grand Cayman in February of 1996. (See photo, pg. 154)

'Lane Taylor Sealy' — Clarence Pleasants especially liked this variety for its large, fragrant, light salmon flowers with pale yellow coronas striped slightly deeper salmon. According to Clarence, it is a prolific bloomer and very striking, even though the flowers do not cover the plants. Newding agrees it is a prolific spring bloomer and a hardy plant.

'Petite Salmon' — Introduced by the Los Angeles State and County Arboretum, this free-blooming variety bears single, bright salmon-pink flowers with yellow coronas. One of the first introductions to bloom continuously in warm climate areas, it is ideally suited to container culture and also makes an excellent low hedge. (See photo, pg. 48)

'Turner's Carnival'™ — (Plant patent #6339) Hybridized by Ted Turner, Sr. who chose the name because, in his words, it is a "circus" of colors. Single, salmon-pink flowers are streaked with dark pink and the curled petals are outlined with burgundy. The plant is everblooming, a small grower easily kept to 3 feet in height and, according to Ted Turner, Sr., is the most "petite" of all oleanders to date. It is excellent planted as a low hedge or as a single, colorful highlight in the garden and is an ideal pot plant. (See photo, pg. 70)

Double Salmon Flowers

'Mrs. F. Roeding' — In a letter to Clarence Pleasants in July, 1981, Mr. George C. Roeding, Jr., of the California Nursery Company, relates the origin of Mrs. F. Roeding:

> I am enclosing a photocopy of an old catalog which indicates that the "Mrs. F. Roeding" was originated by·my father in 1905. Our nursery, at that time, was known as Fancher Creek Nurseries which became part of the nurseries known as and owned today by the California Nursery Company. The "Mrs. F. Roeding" variety is planted along many of our highways in California and is also used for the home. As you may know, it blooms quite profusely and is more compact than most of the oleander grown today.

The 1905 Fancher Creek catalog gives a commendable description:

> This magnificent double Oleander, originated by us, is a chance seedling out of several thousand raised from the imported varieties. [Clarence Pleasants stated that this was one of a group of Japanese seeds sent to the above nursery.] If properly pruned to one stem, the branches form a fine, compact, dense head, covered in summer with trusses of

Petite Salmon

48

beautiful double pink flowers, (the color of the La France rose) delightfully fragrant, with fringed petals which completely envelope the plant.
More hardy than any other variety and is in every respect a very superior Oleander; worthy of a place in every garden.

The large flowers are salmon-pink deepening to almost salmon-orange later in the
season during the hottest weather. 'Mrs. F. Roeding' prefers full sun but is one of the
few plants that will do well with only morning sun. (See photo, pg. ii)

'Mrs. Lucille Hutchings' — A large, hardy plant on the scale of 'Ed Barr' and
'Mathilde Ferrier', this variety makes a fine screen. It requires plenty of room but
little care. Not too attractive when out of bloom but in flower it is truly spectacular
with large blossoms of a delicate peach shade and the tips of the petals tinged pink.
(See photo, pg. 34)

Yellow Flowers

Single Yellow Flowers

'Centennial' — Newding remarks, "Actually, it looks like a fried egg! It has
an orange center and when it fades a little it looks white, but it's really a light creamy
yellow with a very decided orange center. It was named Centennial in commemora-
tion of the 1991 centennial of the University of Texas Medical Branch whose colors
are orange — bright orange." The corona is distinctly striped and plants are half-
hardy. (See photo, pg. 139)

'I. Lovenberg' — Very pleasing, creamy yellow, star-shaped flowers that don't
open all the way. According to Bob Newding it is the same as the variety named 'Isle
of Capri' introduced by Monrovia Nursery. This half-hardy plant can get quite large
but is extremely attractive, flowering all summer. (See photo, pg. 4)

'Isle of Capri' — (See 'I. Lovenberg', above)

'Turner's Kim Bell'™ — A free-blooming variety that has a very long blooming season. Large buds open to beautiful yellow flowers with curved corolla lobes
and a bright yellow corona striped with red. The blossom carries a rose potpourri
fragrance. A Turner hybrid named after Kimberly Bell of Corpus Christi, Texas. (See
photo, pg. 98)

'Sue Hawley Oakes' — A lovely plant bearing single, creamy medium-yellow,
bell-shaped blooms with yellow throats. Flowers appear not to open completely
and are described by some as vaguely star-shaped. Leaves are dark green with slightly
cupped edges. Plants are intermediate in size and somewhat tender but highly prized
for their ornamental qualities. (See photo, pg. 22)

'Turner's Shari D.'™ (Patent #5378) — The first of the Turner hybrids that is
truly a breakthrough, named after Ted Sr.'s daughter Shari Delane. The Turners offer
the following description: "Free blooming, single flowers, a soft buff yellow color
with tinges of pink. Blossom clusters are extremely large and, in fact, at a distance

resemble Rhododendron blossoms. Growth is full and upright. Good for hedges, garden color, small multi-trunk trees and containers. Blooms April 1st through first frost; but will bloom year round with mild winters." Although Shari D. prefers full sun it will also bloom in part shade, requiring only morning sun to flower. Growth is in the 6 to 8 foot range but can be kept at 5 feet with pruning. Plants are half-hardy according to Newding. It is also noted by many growers, including the Turners, that the flowers change color with the day and the season. Cloudy days tend to bring out significantly more yellow and with the sun and heat of summer flowers tend to fade almost to white. (See photo, pg. 146)

'Wimcrest' — A rare plant discovered by Oleander Society member Elizabeth Head on Wimcrest Street in Galveston, Texas. The original plant was destroyed but has been propagated by Bob Newding and others and is now a part of the International Oleander Society's Reference Collection. Newding describes the large, single flowers as having light yellow petals and prominent, medium yellow centers. Medium large, reasonably hardy plants that rebloom. Deep green leaves have occasional yellow splotches. Should be propagated for planting around the island.

Double and Semi-Double (Hose-in-Hose) Yellow Flowers

'Mathilde Ferrier' — A very large, hardy plant that is excellent as a small tree, either trained as a standard or multi-trunk. When planted in the ground as a shrub it needs a lot of room. The double flowers are a soft yellow and very attractive. Perhaps the most common double yellow in Galveston and although often neglected, it responds well to annual pruning and feeding. (See photo, pg. 76)

Harriet Newding

'Sorrento' — Newding calls it probably the most tender of all the oleanders: "I don't know of any that burns quicker in the wind or is more susceptible to freeze, but it has an absolutely gorgeous, soft, creamy, almost lemon sherbet-colored bloom. It's very delicate, very beautiful." This is an excellent, free-blooming plant for container culture and can easily be kept compact. Our photo is of Bob Newding standing by 'Sorrento' and 'Hawaii'. Both plants were flowering profusely one block from the sea at Gaido's Motel. They were planted against a masonry wall that reflects light and absorbs heat during the day releasing it at night but, equally important, blocks the gulf wind preventing desiccation of the leaves. (See photo, back cover)

White Flowers

Single White Flowers

'Casablanca'™ 'Monca' — A Monrovia Nursery introduction from northern Africa in 1978. Compact and free-blooming, plants bear clusters of single, pure white flowers with maroon sepals that appear almost lavender at times. Growth in the ground is moderately fast; height is 8 to 10 feet with about the same width. (See photo, pg. 57)

'Ed Barr' — A big plant that will quickly form a screen around tennis courts or the perimeter of a yard, or hide unsightly areas. An exceptionally hardy variety that bears prodigious quantities of large, white flowers with yellow coronas and remains in bloom most of the summer. This cultivar can get to 18 feet in height with a 10 foot spread if left unpruned but with yearly pruning can be kept to 10 to 12 feet by about 6 feet. (See photo, pg. 58)

'Hardy White' — A name of no recognized standing. Plants listed under this name by nurseries are most commonly 'Sister Agnes' or 'Ed Barr'.

'Harriet Newding' — When we first saw this flower at the Oleander Festival in 1994 we were very impressed with its uniqueness. A seedling that Bob Newding found growing amongst two varieties, 'Hardy Red' and 'Ed Barr,' the plant and flower appear to have characteristics of both — the shape is exactly the same as 'Hardy Red' but the flowers are a parchment white with red sepals and a red stripe extending from the corona through the center of the petals. It is named in honor of Bob Newding's mother. (See photo, pg. 50)

'Morocco'™ 'Monte' — A special selection from seedlings grown at Monrovia Nursery and introduced in 1994. The shape is mounded and according to Monrovia, grows only to a height of 5 to 7 feet. The bright white flowers are borne abundantly for much of the year in warmer climates. (See photo, pg. 56)

'Mrs. Willard Cooke' — A very hardy plant that doesn't get as large as 'Ed Barr' and bears masses of blooms from spring through most of the summer. Large, white, pinwheel flowers develop a streaked red corona as the growing season continues. It is planted on the esplanade in Galveston on I-45. Pleasants said it might be classified as the first plant to bloom because even when it is still chilly you will see blossoms appearing in a warm spot, usually near the ground. (See photos, pgs. 64, 65)

'Mrs. Kelso' — Single, pure white, star-shaped flowers with corolla edges curved slightly upwards are borne on intermediate-sized plants. A spring bloomer; half hardy.

'Mrs. Moody' — Plants known by this name may not be the original 'Mrs. Moody' according to Bob Newding, but a similar variety. He describes it as a pure white flower with faint pink stripes in the corona.

'Sister Agnes' — A beautiful free-flowering variety that is not quite as hardy as some of the other whites but bears great masses of blooms and does exceptionally well on Galveston Island. The single, white flowers

Turner's Kathryn Childers™ Showing flowers on present and previous year's inflorescences

are slightly larger than those of 'Ed Barr.' A grower we know in Florida says it is commonly cultivated throughout the state and one of the easiest plants to grow. (See photo, pg. 74)

'Turner's Daisy'™ — So named because the open-faced flowers in loose clusters remind one of daisies. Borne profusely from April to the first frost, flowers are creamy-beige to yellow edged with blush pink; the yellow corona is striped red. Plants are petite to dwarf and make a fine hedge or a lovely small standard with a weeping, almost oriental effect. (See photo, pg. 102)

'Turner's Elaine Turner'™ — A free-blooming, extremely petite and compact plant with large blossoms of luminous ivory edged with blush pink and yellow, candy-striped coronas. Blooms in loose clusters year round in mild climates and from spring until frost in Zones 8 and 9. Hybridized by Ted L. Turner, Sr. and named for his loving wife Elaine. (See photo, pg. 33)

'**Turner's Kathryn Childers**'™ — (Plant patent pending 5/1/90) A lovely cultivar created by Ted Turner to honor a television celebrity and close friend from Corpus Christi, Texas. Kathryn Childers has led a fascinating life and, according to the Turners, recently retired to devote herself to writing. While in the Secret Service in her 20's, Ms. Childers was assigned to President Kennedy's children and later to President Johnson's children. The plant named for her is free-blooming with soft white flowers blushed pale pink. Flowers do not bloom in clusters but rather cover the entire plant. Full and upright growth makes it excellent for hedges and multi-trunk trees though it also does well in containers. 'Kathryn Childers' blooms all summer and in mild winters will flower on the previous year's inflorescences. (See photos, pg. 52 & 145)

'**Turner's Sissy King**'™ — A free-blooming, dwarf variety with flowers in tight clusters. Individual blossoms have petals of buff ivory blushed with pink and candy-striped coronas. In Corpus Christi, it blooms from April 1st to frost but will flower year-round in milder climates. Hybridized by Ted Turner, Sr. and named for a close friend Mrs. Richard King III. (See photo, pg. 30)

Double and Semi-Double (Hose-in-Hose) White Flowers

'**Magnolia Willis Sealy**' — This is Bob Newding's #1 choice for a semi-double white because, in his words, ". . . it hits all the high points of an oleander; very attractive flowers, relatively hardy, blooms all summer and is fragrant and dependable." Pleasants also listed it as his top selection for a double white. "If I lived in a colder area I would plant 'Ed Barr' first," he says, "I would classify it as hardier. 'Magnolia Willis Sealy' blooms even heavier than 'Ed Barr' and it blossoms from the spring to the cool fall. 'Magnolia Willis Sealy' has these great big, huge clusters of white flowers, very fragrant and a good tub plant." (See photo, pg. 118)

'**Mrs. Knox**' — A large plant that bears hose-in-hose flowers of the purest white with maroon calyces and rounded corolla lobes. The stems of the cymes are distinctly maroon as well. Plants are half-hardy and flower heavily in the spring.

The world's senseless beauty mirrors God's delight.
That rapture's smile is secret everywhere;
It flows in the wind's breath, in the tree's sap,
Its hued magnificence blooms in leaves and flowers.
Sri Aurobindo,
Savitri

Turner's Flirt™

6

New & Dwarf Varieties

The Turner Hybrids

I spent a wonderful day with Ted Turner, Sr. and Ted Turner, Jr. at their nursery Turners Gardenland in Corpus Christi, Texas, recording their experiences with oleanders. We walked through greenhouses filled with oleanders in bloom, photographed Ted Sr.'s latest hybrids, viewed rows of container-grown oleanders being pruned to multi-trunk specimens and patio trees, and visited landscapes that featured their hybrids in full bloom.

Ted Sr. began working with oleanders more than 15 years ago. He was motivated and inspired by two factors; the first was that everyone seemed to be hybridizing hibiscus in those days and no one to his knowledge was working on oleanders, a plant that he loved for its floriferousness and ease of culture. The second occurred on a vacation in the Caribbean islands. Ted Sr. and Ted Jr. were snorkeling in an area full of sea urchins and soon became bored. They decided to explore the dry land instead and while walking around came upon an oleander hedge that was, in their words, "fantastic." It was a dwarf plant with many buds but as Ted Sr. recalled, ". . . some of them [the buds] didn't open up, the blossoms didn't open up. It looked like little pink globes sitting on top of it. Really odd!" They collected seeds from this hedge and their hybridizing program was launched.

With a clear idea of their goal and what they wanted to achieve, Ted Sr. stated: "We already had one of the best oleanders in the world with 'Petite Salmon'. What we wanted were more dwarf oleanders, more colors in a 'petite' oleander and what I call a 'free-blooming oleander.'" Ted Jr.'s definition of a free-blooming oleander is one that will bloom eight or nine months of the year in the Corpus Christi area: "If they're growing, they're blooming, that's what it amounts to. It has to be hot enough for them to grow and if they're growing they're going to bloom outside from April to October or November." Additional goals were to develop plants especially suited for container culture, smaller landscape plants and a dwarf red form.

(See Chapter 5 for detailed descriptions of the Turner hybrids.)

The Monrovia Hybrids

The Monrovia introductions are not only valuable for their smaller size but for their exceptional flower production. Newding reports that in the Gulf Coast area most of them will remain in bloom from April through the hottest part of the summer until the first cold snap which usually occurs in November. Another outstanding characteristic is their bright, crisp color quality which Newding describes as "iridescent" and "neon." As he says, "In the early morning sun and the twilight they almost seem to glow."

Ted Turner, Jr. also remarked on the flower quality of the Monrovia series calling the red of 'Algiers' "fluorescent" and noting the tendency of several of the varieties to bloom more profusely in the fall as does 'Turner's Tickled Pink'. This is especially useful in landscaping for providing color at a time of year when many other plants have finished flowering.

Dwarf Series — 'Marrakesh'™ and 'Morocco'™. The origin of these two varieties was seed collected in 1986 by Audrey Teasdale, a botanist at Monrovia Nursery. The seedlings were grown at Monrovia's Azusa, California, location and 'Marrakesh' and 'Morocco' were selected in 1989. The goal of this selection process was to develop a series of oleanders that were small in stature and better suited to today's smaller gardens. The series was released to the nursery trade in 1994.

Intermediate-sized Series — 'Algiers'™, 'Casablanca'™, 'Tangier'™. While in Algeria, former chairman of the board of Monrovia Nursery, Howard Past spotted these colorful, intermediate-sized specimens in a cemetery. Recognizing their commercial potential with the main benefit being reduced maintenance and adaptability to average

Photo courtesy Monrovia Nursery Company

Morocco™

residential landscaping, he took cuttings back to California where they were rooted. The series was introduced by Monrovia in 1978.

Intermediate Size — 'Ruby Lace'™. This variety was given to Monrovia Nursery in 1984 by a hobbyist who had been growing the plant in his yard in Anaheim, California, for 19 years. It was then propagated by Monrovia and introduced in 1986.

(See Chapter 5 for detailed descriptions of the Monrovia hybrids.)

The Israel Hybrids

For some years Clarence Pleasants corresponded with Mr. Zafrir, the author of *Oleanders In Israel* and sold copies of Mr. Zafrir's book in the United States. Through their correspondence, Clarence received some cuttings of dwarf varieties but they arrived in very poor condition and none survived. Another time an Egyptian gentleman visiting the United States told Clarence of the hybridizing work in Israel and mentioned that many varieties were carefully selected for their exceptionally dwarf characteristics and their hardiness in the desert.

Photo courtesy Monrovia Nursery Company

Casablanca™

To create a little flower
Is the labour of ages.
William Blake

Ed Barr

Ed Barr

7

It is through flowers that Nature
expresses herself most harmoniously.
The Mother

Oleanders and Other Poisonous Plants

We would like to preface this section with observations by a number of members of the International Oleander Society concerning the poisonous qualities of oleanders. It is their contention that it has suffered bad press and they hope to rectify this with factual data. Kewpie Gaido called it her "holy grail" to bring to the attention of the public the fact that although millions of oleanders are grown in southern Texas and on Galveston Island, there have been no reports of fatalities due to the ingestion of oleander leaves for many years. In fact, as far as we have been able to ascertain from various newspaper articles, fatalities from all poisonous plants are rare. (Poisonous mushrooms are an exception; as of this writing there is a recent report of the death of a child from ingestion of Death Cap mushrooms.) According to Gary Outenreath of Moody Gardens, there have been no accidental deaths attributed to the ingestion of poisonous plants since the early 1970's. He echoes Kewpie's sentiments by pointing out, ". . . the most unusual thing is that they [oleanders] are somewhat maligned in terms of the perception of toxicity. While they are very toxic, the Castor Bean is the third most toxic substance known to man and there is no antidote for it." Gary also mentions that azaleas and English Ivy are much more poisonous than oleanders. In an interview with Elizabeth Head in 1995 she noted that in a recent article on poisonous plants in the Galveston newspaper, oleanders were not listed among the top ten most poisonous. According to the Spring '95 issue of *Nerium News,* the ten plants that most frequently cause human poisonings are: Philodendron, Pepper, Dumb Cane (Dieffenbachia), Poinsettia, Holly, Pokeweed, Peace Lily, Jade Plant, Pothos (Devil's Ivy) and Poison Ivy.

Oleanders are widely grown throughout Florida, especially in seaside areas and traffic medians. In California hundreds of miles or freeways are planted with oleanders and they are featured plants at Disneyland, Disney World, Cypress Gardens, Moody Gardens and many other public parks and gardens. That they are not considered particularly risky to the public in such accessible locations may be due in part to a fact that was constantly mentioned in discussions of their potential danger. The sap of oleanders, a clear, sticky substance, is extremely bitter and also a natural emetic so the smallest taste would likely cause a child, or an adult, to spit it out immediately. In the movie *Dragonwyck*, Vincent Price poisons his wife by substituting ground oleander for herbs. This is undoubtedly an exaggeration since the taste would have been so horrible she probably would not have been able to swallow the first bite. We have read of people taking only one bite, immediately spitting it out and having the bitter, unpleasant taste remain with them for hours. As testimony to its poisonous quality, there are accounts of primitive tribes using oleanders as a way of administering justice.

Numerous plants that are known to be highly toxic are routinely cultivated as ornamentals in the United States. Many members of the Euphorbiaceae family are poisonous to some degree; the sap of some may cause a dermatitis similar to poison ivy, others are so toxic the sap is used to tip poison arrows. Poinsettias, poisonous members of the same family, are grown by the millions to decorate homes during the Christmas holidays, and the Pencil Cactus, *Euphorbia tirucalli*, caused widespread blindness among Mussolini's unsuspecting troops in north Africa during W.W. II when, after marching through fields of the plants, they rubbed their eyes. The Dumb Cane, *Dieffenbachia*, is highly poisonous and yet remains one of the most popular indoor plants. One could add to this list *Plumerias*, *Lantanas*, many forms of *Solanum*, especially the Nightshade, and as mentioned above, *Ricinus*, the Castor Bean.

Nevertheless, all parts of the oleander plant — roots, stems, bark, leaves (both fresh and dried) and flowers — are poisonous to some degree and there have been reports of deaths from ingestion as far back as Theophrastus who mentioned the poisoning of animals during the campaign of Alexander the Great (334-323 B.C.). Animals, usually possessed of a sixth sense about which plants are poisonous, tend to stay away from them. (We would perhaps be remiss, however, not to offer these examples of the inconsistency of animals' instincts regarding poisonous plants: Our beautiful white Persian cat munched our Christmas poinsettias throughout the holidays with, surprisingly, no obvious ill effects, and Bob Newding's Muscovy duck follows him around as he prunes his oleanders, continuously eating and spitting out the leaves as they fall to the ground.) On Hilton Head Island, South Carolina, oleanders are considered the only plant safe from the overabundant deer population but there are, nonetheless, numerous accounts of animals being poisoned. F.J.J. Pagen states in his book that 15-20 g of fresh leaves is sufficient to kill a horse and only 1-5 g will kill a sheep. In Italy, the oleander is commonly called *Ammazza l'Asino*, 'Donkey Killer'; in Persian, it is *Khar Zahr*, 'Donkey Poison'. It is recorded that livestock has perished when fed with hay in which oleander leaves were mixed so it is a good practice to keep oleander plants away from all hay and animal feedstocks as well as grazing areas.

Most of the recorded instances of human deaths are from the 1700's and 1800's with a few cases reported in the 1900's. There is no doubt, however, that oleanders *can* cause death. The symptoms of oleander poisoning are nausea, vomiting, dizziness, dilation of the pupils, bloody diarrhea, cardiac weakness and coma preceding death. Most instances of recorded poisonings have been attributed to using the branches as skewers for meat or hot-dogs. In Hawaii children are taught at an early age never to use oleander branches for barbecuing or toasting marshmallows. Oleander wood should not be used for firewood as the smoke can cause severe and painful irritation. Precautions are well worth taking when working with oleanders as the clear, sticky sap may cause skin irritation for people with sensitive skin. The use of normal prudence when handling poisonous plants, even if they are something as familiar as poinsettias, is advised. Do not rub your eyes with your hands while working with poisonous plants and always wash thoroughly with soap and water after handling them.

Although the above warnings should be taken very seriously, one finds certain authors who have gone to extremes by warning that the presence of oleander flowers themselves can cause sickness in an enclosed area.

It is important to become familiar with the various kinds of poisonous plants in your area. Some are wild but many are favorites in our homes and gardens. Teach children, the most frequent victims, to recognize and avoid them. Listed below are some of the most common poisonous plants <u>with the potential to cause fatality</u>. Many others have less extreme but still serious effects such as restricted breathing or respiratory paralysis, intense pain, diarrhea, vomiting, convulsions, permanent corneal damage, heart damage, coma, etc. For more information on poisonous plants and their symptoms, call or write the local Poison Control Center in your state.

Common and Botanical Name	Toxic Parts	Comments
Azalea *Rhododendron spp.*	All parts	One of America's favorite plants.
Carolina Jessamine *Gelsemium sempervirens*	All parts	Lovely and fragrant, flowers are highly toxic.
Castor Bean *Ricinus communis*	Seeds	Three seeds can be fatal.
Cherry Laurel *Prunus caroliniana*	All parts	Contains cyanide-producing substances.
Chinaberry *Melia azedarach*	Entire Plant	Trees fruit heavily and berries have caused many fatalities in children.
Horse Nettle *Solanum carolinense*	All parts	Death from paralysis can occur from the ingestion of small amounts of this Nightshade.

Hydrangea *Hydrangea spp.*	All parts	A number of species contain the toxin glucoside.
Jimsonweed *Datura stramonium*	All parts	<u>Very small amounts</u> have proved fatal.
Lantana *Lantana spp.*	All parts	Small amounts can cause coma.
Larkspur *Delphinium spp.*	Seeds, new leaves	Cause of <u>many</u> deaths in cattle.
Mistletoe *Phoradendron serotinum*	Entire plant	
Virginia Creeper *Parthenocissus quinquefolia*	Entire plant	Death can occur due to tetanus.
Yellow Oleander *Thevetia peruviana*	Entire plant	Fleshy outer portion of seed is used in parts of Asia to commit suicide.

To him who in the love of Nature holds Communion
with her visible forms,
she speaks a various language.
William Cullen Bryant

8

What a joy life is when you have made a close working partnership with Nature.

Luther Burbank

Oleander Culture

To begin this rather lengthy chapter which attempts to detail all aspects of oleander culture, we would like to offer a thought from Kewpie Gaido about her beloved oleanders. "They are so beautiful," she said, "and with just a little care they do well and show their appreciation by doing better than they had originally been known to."

*All the flowers of all the tomorrows
Are in the seeds of today and yesterday.*

Chinese proverb

Propagation

Seeds

Seed pods should be collected in the United States in warm climate zones during late December and early January. In the tropics they ripen throughout the year but are most abundant from winter to early spring. Keep the pod in a paper or plastic bag or container until it splits open to reveal the seeds. Oleander seeds are highly viable and germination can be nearly 100% with many cultivars. Seeds are very small and thin, usually not longer than ¼ to ½ inch or even smaller, and have a coma at one end.

Planting Media for Seeds

Always use a sterilized planting medium that is loose and provides good drainage but is sufficiently moisture retentive. There are a number of commercial mixtures that are excellent. These are basically lightweight "soilless" mixes composed primarily of finely ground spaghnum peat moss mixed with perlite (occasionally vermiculite) and, in the best products, a surfactant, or wetting agent. Some formulations also contain a "starter charge" of diluted, water soluble nutrients, not strong enough to burn but sufficient to get the germinating seedlings off to a good start.

Planting the Seeds

Moisten planting media thoroughly prior to sowing the seeds. It is helpful to level the media in pots or trays with a flat object such as a small block of wood. This assures uniform water distribution and makes it easier to cover the seeds evenly. After placing the seeds on the media cover them very lightly, no more than ˘ inch deep, preferably ⅛ inch. We refer to this as the "salt and pepper treatment" and will often sift our planting media through a fine mesh in order to provide the finest covering material. After covering the seeds tamp the soil with the same block of wood used for leveling. This will assure good soil contact with the seed and promote higher germination.

In our plumeria handbook we mentioned planting seeds vertically with only the papery wing exposed above the soil line. Oleander seeds may also be planted in this manner by inserting the tip of the seed into the media so that only the coma (tuft of hairs at one end) remains above the soil level.

Watering Seeds and Seedlings

Bottom watering is always preferred whenever possible but since oleander seeds are not minute and dust-like, a watering can with a fine rose to prevent the media from washing away will do very well. Keep the soil damp, not wet, until the seeds germinate. Overwatering and planting seeds too close together can lead to damp-off, a fungus that can destroy a crop of seedlings in an amazingly short time. There are commercial fungicides one can apply to

Mrs. Willard Cooke

prevent or retard damp-off but it is always best not to plant seeds too closely and never overwater.

Germination

Success with seeds depends on a number of factors but these are not complicated. Most importantly, seeds must be fresh. Oleander seeds do have a very high viability, as mentioned above, but they must be sown within a few months of harvesting or viability drops off considerably. Year old seeds are generally no longer viable. Soil temperature is also critical. A cold, damp soil is more likely to cause seeds to rot than allow them to germinate. Either plant in warm weather, or supply bottom heat with a heating mat or heat coils kept at a temperature of 75 to 80 degrees. Germination usually takes place in ten to twelve days and initial growth is fairly rapid. As a further indication of the viability of oleander seeds, we recount Dr. Parsons' experience with seed that was supplied by Elizabeth Head from the finest cultivars

Mrs. Willard Cooke

in Galveston and irradiated at Texas A&M University. Dr. Parsons said, "Anyway, I had them irradiated and they [Texas A&M] sent me the seed back. I don't know if you've seen oleander seed, they look like cotton. And I said, these damned things are not going to come up and the next thing I know they're as thick as hair on a dog's back and I end up with 600 seedlings!"

Care of Seedlings After Germination

As seedlings germinate, apply a dilute solution (usually ¼ strength) of a balanced water soluble fertilizer such as 20-20-20. The first set of leaves are called cotyledons, or "seed leaves," and are generally different in appearance from the "true leaves." When the first true leaves appear, increase the fertilizer to half-strength. When seedlings are about 3 to 4 inches high it is time to transplant them into individual pots.

Pros and Cons of Propagation by Seed

One of the greatest pleasures of gardening is raising a plant from seed and following its growth and development through to the moment it flowers. An important benefit of planting oleanders from seed is that one can, with very little expense, raise a great number of plants and with reasonable certainty, depending on the quality of the original cross, find a new color, shape, form or fragrance in a large batch of seedlings. In addition, one may be fortunate to develop a plant with greater hardiness or disease resistance. It must be kept in mind, however, that most seedlings will be inferior to the original parent plants and should be destroyed after evaluation. Propagating from cuttings assures that the new plant will be identical to the original but only with seeds can we discover something new.

An Additional Note on Seedlings

In general, it takes from one to two years for an oleander to flower from seed. After planting seeds in spring, pinch back seedlings in late summer or early fall to encourage branching. If not pinched out many seedlings tend to grow straight up.

Mrs. Runge

Most plants, if started from seed in April or May, will go through one growing season and flower in late spring of the following year.

Cuttings

Oleander cuttings are very easy to root, even for beginners. Cuttings should be taken in early spring when plants are growing vigorously as they will root most quickly at that time. Also, unless one has access to a greenhouse with a cooling system, hot summer temperatures elevate bacterial metabolism in cuttings so they will tend to rot before they can root. Donald J. Moore, former Superintendent of the Botanical Gardens in Bermuda, mentioned that cuttings there are best rooted in nursery beds or even planted directly into the ground where they are to grow. This was also our experience in the tropical climate of southern India. There we developed a rooting media of coir dust, the granular residue of the coconut husk that remains after the fiber has been extracted, mixed with sharp sand. We placed cuttings in the shade of a mango tree, out of the wind, and consistantly had close to 100% rooting. Kewpie Gaido tells the story of her father's experience with an oleander. He needed to stake a special plant he had just acquired and since there were oleanders all around, cut a long straight oleander stem for a support. In short, his plant died and the oleander rooted!

In his experience with variegated oleanders, Dr. Jerry Parsons discovered that, like many other variegated plants, they were more difficult to root whether the variegation occured in the flowers or the leaves. (He also found this to be true of variegated plants propagated by tissue culture.)

Type and Size of Cuttings

Tip cuttings make the most attractive plants so always select a branched cutting and you will have a nicely shaped plant almost immediately. If the tip cutting is not branched, the top may be pinched back as soon as roots have formed and the axillary buds will quickly produce shoots creating a well balanced shape. The best wood is at least a year old and anywhere from $3/8$ to $5/8$ of an inch in diameter. Cuttings up to 24 inches in length may be rooted either in water or in a soil media if they are taken at the proper time when in an active growth stage. Many gardeners in tropical climates also use the suckers that form at the base of plants.

Rooting Media for Cuttings

As we have found with plumerias, one of the best mediums for rooting oleanders is coarse perlite, or a mixture of three parts coarse perlite and one part peat moss. This assures good drainage and aeration. In recent years, Elizabeth Head of the Oleander Society has made numerous tests of rooting media and has also concluded that the highest rooting percentage has been with coarse perlite. As we observed with plumerias, the newly emerging roots attach themselves to the perlite granules and form a mass that holds together well during transplanting. We have also used sharp sand (builder's sand or concrete sand) with excellent results. Do not use mortar sand as it has none of the sharp particles of silica that improve aeration and drainage. Some growers mix a very small quantity of peat moss into the sand (not more than 5 or 10%) for added moisture retention.

Rooting Cuttings in Media

This is the simplest and most effective way to root large or small quantities of oleander cuttings. Keep media moist but not wet, and do not allow it to dry out at any time during the rooting period. As soon as you see new leaves emerging, begin fertilizing with a water soluble fertilizer such as the 20-20-20 formula recommended for seedlings diluted to ¼ strength. Resist the temptation to use full strength fertilizer as it will burn the newly emerging roots. Rooted cuttings will grow very rapidly with supplemental fertilizing. Transplanting may be done at almost any stage when new growth is evident.

I asked the Turners, who have rooted thousands upon thousands of cuttings, what problems, if any, they had experienced. Ted Sr. offered some illuminating advice. Having had some losses due to a fungus which especially affected the petites, they began dipping the cuttings in a mixture of two tablespoons Captan per gallon of water before placing them in the rooting media. "I prefer getting cuttings in between rains and not right after a rain when they're full of moisture," he stated. "We dip them, just drench them [in the Captan mix] and then let them dry almost completely. And then I use Rootone. We stick the base of the cutting and up about ½ inch of stem into the Rootone. I still use Rootone because my greenhouse is not a propagation house so I have more trouble now than I used to. We used to have mist and we could just put them in any way we wanted to and they would root. But I have a mixture, you saw what I have, there's every kind of plant imaginable in that greenhouse, so I have more fungus in there from having to use too much moisture on some of the tropicals and oleanders don't like that. They like a hot dry house where most of the tropical plants like a humid greenhouse." It is important to note that if you have a greenhouse without mist, take cuttings of hardened wood. If you have a mist system, take transitional (half hardened) wood rather than the soft, green wood of new growth.

Inserting Cuttings in Media

Make a cut, using sharp, sterile pruners or pruning knife, at the base of the stem just below a node. Cut on a 45-degree angle to provide a larger surface area for roots to form. Moisten the cutting, shake off any excess water and then apply a rooting hormone. (It is a good idea to put a small quantity of rooting hormone into a separate container so as not to contaminate the remainder with moisture.) Bob Newding shares a special technique he uses to encourage more roots. At a tangential angle to the stem, make a cut just under the cambium 1 to 1½ inches up from the base of the cutting. Open the flap slightly and with the knife blade insert some Rootone or other rooting hormone under the flap. This will encourage roots to form throughout the area of the cut and not just at the nodes immediately above the base of the cutting. The extra work is worth the effort, especially on rare plants or those that have proven somewhat difficult to root. Before inserting the cutting into the media, make a hole with a stick or another cutting of equal size to the required depth so that when you insert the cutting the rooting hormone remains undisturbed. Larger cuttings should be inserted 3 to 4 inches or more into the media so they will be supported, smaller cuttings up to 2 or 3 inches. It is helpful to place them at a 45 degree angle as this allows easier transport of nutrients to the tip of the cutting during the rooting process. Water-soluble rooting hormones such as Dip 'N Gro are commercially available, very easy to use, and are absorbed more readily by the tissues of the cutting. (Be sure to use distilled water when preparing a solution.) We have had excellent results with rooting many types of plants with Dip 'N Gro but have not tried it on oleanders as yet. We have no reports of anyone using water-soluble rooting hormones to date and would appreciate receiving any information for inclusion in future editions of the handbook.

Turner's Tickled Pink™

Sun/Shade Requirements for Rooting Cuttings

Bright shade, filtered light, a north-facing windowsill or a bright room out of direct sun all work well for rooting cuttings. The most important thing is to keep them out of the wind and cold.

Transplanting Cuttings Rooted in Media

The roots of cuttings rooted in sharp sand or other media are strong and transplanting is easily accomplished using normal caution and standard techniques.

Ted Turner, Sr. with Turner's Carnival™

Comparison of Turner's Carnival™ (left) and Turner's Tickled Pink™ (right)

Extra care may be necessary when lifting cuttings from sand, however, as the weight of the sand may cause some breakage.

Rooting Cuttings in Water

Several members of the Oleander Society find this to be the easiest method of rooting cuttings. The only drawback is that roots formed in water are very brittle and have a tendency to break easily when transplanted. Choose a branched cutting from 10 to 18 inches long and remove all but the new terminal leaves. It is important to check that the plant is vigorous and healthy and free of insects and disease, especially bacterial gall. Use a non-metallic container such as glass or plastic and check the water level regularly, being sure not to allow algae to develop. Depending on the variety, cuttings will root in five to twelve days but if taken too early in the spring may take up to three weeks. Some growers suggest adding a very minute quantity of fertilizer and rooting hormone to the water; others feel this is unnecessary. Cuttings taken with a "heel" will assure an even higher percentage of rooting in water (or media) but the extra work is hardly warranted since the rooting percentage will almost always be over 90%.

We have read published recommendations for rooting large stems 4 to 6 feet tall in water. This is erroneous information and will rarely succeed. Biologist Bob Newding explains why: "The reason you cannot do it is that the plant will rot before it will root. You are looking at cells that are trying to metabolize without any roots and they generate heat. The bacteria grows very quickly in warm water — because you need warm conditions to do it — and the darn things rot."

Transplanting Cuttings Rooted in Water

As mentioned above, the roots of cuttings rooted in water are extremely brittle and unless handled with the greatest care will snap off easily. Transplant rooted cuttings when approximately four to six roots have emerged using the following technique. Select the proper size container consistent with the size of the cutting. Using a very light, highly organic potting soil, fill the container so that the cutting will be at the proper depth, mounding the soil slightly in the center. Lift the cutting from the water and place it carefully on top of the potting soil in the container. Do not push the roots into the soil but gently spread them on top of the mound. Slowly fill the container with more soil to the correct level. Do not pack the soil around the roots! Water gently and the soil will fill the interstices, eliminating air pockets. If you are concerned about breaking roots, allow the cutting to remain in the container until it is filled with roots. At this point the roots will be hardened off, are no longer so brittle and can easily be transplanted.

Air Layering

Air layering is an excellent technique for propagating difficult-to-root plants plant. Most of the time oleanders root so easily in water or media, and so much more rapidly, that air layering is not required. There are occasions, however, when air-layering is recommended. Rare and one-of-a-kind plants, or particularly challenging plants, are best air-layered to be absolutely sure of success. It is also the best method when you wish to start a tree-form specimen. For those seeking instant gratification, choose a branch with latent buds and you will have a blooming plant in a matter of weeks.

For a tree-form plant, make the layering at a point on a branch so that the portion above the layering (that will be the new plant after roots form) has the desired length and girth. For a ready-made bush shape, make the layering just below a section where there are already several branches. To make the layering, cut upwards into the stem at a 45-degree angle with a sharp, sterile knife or blade, stopping about one third of the way through. Place a tiny pebble or gravel chip in the cut to keep it from closing and healing over. A powdered or liquid rooting hormone may be applied. Place a double handful of damp sphagnum moss over the entire area, cover with plastic or plastic wrap, and secure at both ends. While you work support the upper portion of the branch so that it does not bend or break at the point of the cut. Sufficient moisture should be retained by the plastic during the time necessary for rooting but protection from hot sun and wind should be provided. (There are numerous "How-To" books available that illustrate all types of layering.)

Ground Layering

Ground layering is another technique for rooting large stems to be trained into tree-forms. Simply bend a stem of the desired height and, as near the ground as possible, make a 45-degree incision on the underside upwards into the stem about one-third of the way. As with air layering, the stem still receives nutrients from the main plant and one is assured of rooting. Again, keep the cut open by inserting a very small stone chip or pebble so that the stem cannot heal around the cut. Use a container filled with a porous rooting mixture as described above for cuttings and be certain the bottom of the stem (where you have made the cut) is fully in contact with the soil. This is not quite as easy as it seems when one is handling large stems and it is sometimes necessary to raise the pot of rooting medium off the ground. Place a heavy stone on top of the container so the stem is held firmly in place and cannot spring up. Monitor the container carefully as it is important that the soil should not dry out.

Grafting

The Turners have grafted the petite oleanders onto 'Hardy Red' with excellent success, creating a tree-form plant with the compact, long-blooming head of the petites. They allowed five feet of trunk and since the head was small, the tree did not require staking even in the high winds of the Gulf Coast. One can also create "rainbow" trees by grafting several selected cultivars onto hardy stock.

Ted Sr. commented, "I read an old, old book that said you can graft the Desert Rose [Adenium] onto oleander and the Desert Rose will bloom twice as good. I haven't done it yet but I'm going to do it." Frank Pagen agrees stating: "*Adenium spp.* can be grafted on a *Nerium oleander* rootstock to produce profusely flowering plants."

There are many types of grafts described by names such as wedge, saddle, side, veneer and cleft, with whip-and-tongue being the most common. The plant with the desirable flowers, fruits and foliage that is to become the new "top" is called the scion. The plant that will provide the root system is called the stock. The success of the graft requires that the cambium (the tissue just beneath the bark) of the scion and the stock is joined so that it eventually grows together. This is accomplished by assuring that the stock and scion are the same thickness. Gardeners who think nothing of sticking a hundred cuttings in an hour sometimes hesitate to try grafting, perhaps thinking it too difficult and unpredictable. It is true that some people seem to have a certain "touch" but with just a little practice anyone can achieve success. There are many books containing illustrations of different types of grafts that are helpful to the beginner and we encourage everyone to give it a try. Fascinating combinations, such as the "rainbow" tree mentioned above, can only be accomplished by grafting.

Soil Preparation for Planting in the Ground

In discussions with growers and horticulturists over the years, the consensus seems to be that oleanders will grow in almost anything! As Ted Turner, Sr. remarked, "They grow in oyster shell and sand, Galveston proves that hands down." Having

been involved with composting and soils for so much of our lives, however, we can't resist offering some general suggestions to encourage the best performance from your oleanders.

Oleanders grow much better in sandy soils than clay soils. If you have a clay soil, amend it with ample amounts of organic matter, preferably compost. Well-decomposed bark, composted hulls or leaf mold are also very useful. Do not use fresh bark or any other raw organic material as nitrogen "draft" will occur and the nutrients meant for your plants will be used by bacteria to break down the organic

General Pershing

73

Sister Agnes

matter. The two most important elements plant roots require are oxygen and drainage. All clay soils require flocculation and adding compost or decomposed organic matter to your planting media is the best way to insure this. We also recommend adding organic matter to pure sand and sandy soils. Even though oleanders may be found growing in such soils, the addition of organic matter will aid moisture retention and will help prevent leaching of nutrients.

Any extra effort when preparing a planting hole will pay huge dividends in the future. This is why we recommend digging a hole at least twice or three times as wide as the root ball of the plant, but no deeper. If the first cause of plant mortality is overwatering, the second is usually planting too deep. By carefully measuring the depth of the root ball one can prepare a hole of the proper depth. One of the guidelines we have always used is to plant the root ball an inch or two above ground level to allow for future settling, then cover the surface with a solid three inches of mulch.

A current school of thought suggests that adding any organic matter to the backfill mix creates an interface problem and prevents a plant from adapting to the soil in which it will grow for the remainder of its life — but we don't subscribe to it. Having worked in laterite soils in India, black gumbo soils in Houston, and now the red clay hills of Georgia where even power augers have difficulty penetrating, we have witnessed the struggle of plants in unamended planting media. For the last forty years we have always added approximately one-third compost, well-decayed leaf mold, or other fully decomposed organic matter to our backfill mix and have always had excellent results. The only exception is the unique situation of highly saline soils. From their experience with salinity in the Gulf Coast area of Texas, the Turners recommend being very careful about adding too much organic material to such soils as excess salts will be retained instead of leached, thereby causing toxic shock to the plant roots.

It is important to understand that while oleanders will grow in poor soil, they will be at their best in well-drained soil with average fertility. But to balance all this technical information, we should, once again, emphasize how easy it is to grow oleanders. As Dr. Jerry Parsons says, "I tell people, if the highway department can grow them, anybody on earth ought to be able to grow them . . . they're tough!"

Planting and Transplanting

Deep in their roots, all flowers keep the light.
Roethke

Basic Guidelines for Transplanting

We'd like to share some basic rules for transplanting that will assure success for first time gardeners as well as experienced hands. Whether you are transplanting seedlings or rooted cuttings, the same rules apply. Before beginning have all your materials ready. Fill pots to appropriate levels with potting soil and have additional potting soil ready for topping up after transplanting. Have water handy and work in the shade, preferably late in the day so plants will have the night to recoup. If that is not possible, then transplant in the early morning while it is cool. To reduce the stress of transplanting we have always advocated the use of Superthrive, a transplant formula with vitamins and hormones, as it has consistently given us excellent results. When I was in my teens I often went with my father to visit his friend and grower Martin Fleischut where I learned many things about plants, but none more important than the fact that the roots of seedlings and cuttings could die in one minute if exposed to heat and wind. I have followed Mr. Fleishut's advice all these years and the moment we prick out a tender seedling or lift a rooted cutting, it goes into its new container immediately, is watered, and set in shade or filtered light where there is no wind.

Planting in the Ground

Oleanders are easily transplanted directly into the ground and quickly form a mass of fibrous roots under the proper conditions. Remember to add a time-release fertilizer at the time of planting. In general, oleander roots move laterally rather than vertically. The exception to this is when they have insufficient water and have to find an additional source of moisture on their own. There have been many instances of oleander roots clogging old clay drainpipes so planting away from drains, sewer lines and septic tank fields is always advisable.

Once again, before removing the plant from its container be certain you have everything in readiness, i.e. prepared backfill mix, water, mulch, etc. On removing the root ball from the container check to be certain there are no encircling roots. If the plant is heavily rootbound the roots will need to be "combed." If encircling roots are very large they should be removed with pruners. Otherwise, like a bad habit, they will continue circling and never change their ways by spreading out or descending into the soil. With the "comb" of your choice, tease the roots down and away so they will spread outward in the hole. Do not cut away too many roots unless a plant is so potbound that there is no alternative. The general guideline should be to disturb the root system as little as possible, keeping plant stress to a minimum. After planting, water in thoroughly to dispel any air pockets in the soil and add transplant vitamin or root stimulator.

There are two basic watering techniques that are effective in getting sufficient water to the root ball. Both are excellent and although we generally prefer the first, we have used the second with equal success. After planting, instead of firming the soil around the root ball, push the end of a garden hose (with just a trickle of water at first) into the soft soil between the edge of the root ball and the edge of the planting hole as deep as it will go. You will soon see the water pulling the soil into and around the rootball. If water comes immediately to the surface, decrease the water pressure and try again in another spot. Move the hose to three sides of the rootball, letting it soak in each area until water begins spilling out. The second method consists of tamping the soil until it is firm, building a berm around the outer area of the root ball and then allowing the hose to trickle on the surface until the berm is filled and the water percolates down. Depending on the size of the hole this may have to be repeated two or three times to insure thorough soaking of the media. The berm may be left in place through the first season as it is useful whenever you need to provide supplemental water.

Mathilde Ferrier

Sun/Shade Requirements

The major factor that influences color in oleanders is sunlight. Most oleanders require full sun, or nearly full sun, to be at their best. In shaded locations plants bloom sparsely and will tend to stretch and become leggy in their attempt to get more sun. There are several varieties, however, that will perform well in part sun. Newding mentions that 'Sorrento' will bloom in partially shaded locations. According to the Turners, 'Shari D' will bloom in 50% shade and it doesn't matter whether it is 50% filtered light all day or full sun for half a day. For most varieties the amount of sun has a direct effect on flower production so it is important to provide as much heat and sun as possible for best performance.

Watering Plants in the Ground

The first year after transplanting is the most critical time in an oleander's (and most other plants) development. During this period it is important to provide supplemental watering (especially if rains are irregular) to encourage development of a solid root mass capable of sustaining the plant for years to come. When you water, water deeply to encourage the roots to travel downward. Frequent light waterings encourage feeder roots to remain at the surface with potentially serious consequences if waterings are missed and extended droughts prevail. After the first year, supplemental watering should still be provided during periods of drought. Although established plants will survive drought, additional watering will keep them in a much healthier condition and help maintain flower production.

Humidity

Although we have seen that oleanders are highly adaptable to all kinds of weather conditions, thriving in dry desert air as well as the high humidity of the Texas Gulf Coast, the Turners observe that there is something different about high humidity in a greenhouse and high humidity outdoors. They readily admit they don't really know what the difference is but have observed that when conditions are too damp in the greenhouse, spent oleander blossoms remain on varieties that are normally self-cleaning. This same phenomenon has also been observed in the hot, humid climate of Malaysia.

Fertilizing

Our recommendation for fertilizing oleanders to promote maximum flowering is to use a formula high in phosphorus. We alternate a controlled-release fertilizer with an NPK (N=nitrogen, P=phosphorus, K=potassium) of 5-32-5 with a water-soluble formulation of 9-58-8. The low levels of nitrogen maintain plant health but do not push growth. For basic maintenance of oleanders growing in sand or sandy soils with little or no organic matter, fertilize twice yearly with a controlled-release fertilizer using a balanced formula such as 13-13-13, alternating with a high phosphorus formula since nutrients tend to leach quickly in light soils. In areas such as Galveston, where soils are already very high in phosphorus, additional amounts will prove detrimental as too much phosphorus binds micronutrients rendering them unavailable to the roots. It is always best to take a soil test to determine the basic nutrient availability, or lack of it, in your soil whether plants are in the ground or in containers.

Timed-release or controlled-release fertilizers are designed to assure a constant supply of nutrients over a given period. However, due to the nature of the timed-release mechanism — the most widely used formulation being a coated prill that releases nutrients through osmotic action based on heat and moisture — the release curve of nutrients is more rapid in hot weather. Under these conditions, nine to twelve month formulas will be expended long before the suggested period for nutrient release. We have found the three to four month formulations to be more effective in warm climate areas with two applications per year, one in early spring and one in summer.

The best technique for fertilizing large established plants is deep root feeding. We recommend the following: For granular, slow-release products make holes approximately 6 to 12 inches deep, spaced 1 to 2 feet apart and placed 1 to 2 feet on either side of the drip line. (The drip line is the area just below the outer branches where rain would fall off the leaves in the greatest concentration. It is also the area where the most feeder roots are located.) The holes can be made with a rock bar or pick, or even a ¾ to 1 inch piece of rebar or pipe, and the work is much easier after a good rain when the soil is moist. Place slow-release fertilizer in the holes according to product specifications then fill the balance of the hole with compost or enriched topsoil. Repeat the process at recommended intervals. Organic gardeners might like to try filling the holes with compost, composted manure, or manure "tea" as we did in India where slow-release fertilizer products were unavailable. The second technique involves the use of a root feeder, an implement readily available at most garden shops. This is basically a hollow tube with a head that contains a chamber for a pre-packaged nutrient and connects to a garden hose. The force of the water makes the hole and then dissolves the nutrients to fill the hole.

Plants grown in containers also flower best with a fertilizer high in phosphorus. Elizabeth Head related one experiment in fertilizing oleanders that deserves mention. She and Kewpie Gaido were preparing for a forthcoming Lawn & Garden Show in the month of February in Houston, Texas. They had many oleanders in containers but none that would be in bloom so early in the season. So they "pushed" the plants with heavy applications of fertilizer and were able to bring them into bloom for the show. Some seedlings that were only a few months old also flowered. Kewpie commented, "I have a bad habit of overfertilizing. I just throw the Osmokote on like it's free. But I don't think you can kill an oleander with too much fertilizer." (Note that she refers here only to a controlled-release fertilizer!)

The addition of bone meal as a slow-release nutrient to promote flowers and build hardier plants is also recommended, as is the application of chelated trace elements if the soil is lacking minerals. Magnesium sulfate (Epsom salt) helps to prevent leaf scorch and being the core of the chlorophyll molecule is a resonator for the photosynthetic process. In our first book *The Handbook On Plumeria Culture*, we discussed experiments with potassium to strengthen cell walls in plumerias, thereby promoting greater cold tolerance. This would be an excellent research project for oleanders. Again, before adding any trace elements to your soil, we strongly recommend taking a soil test with one of the many kits available or having your local agricultural extension office send samples to a lab for testing.

Mulching

As a founder and vice-president of Living Earth Technology of Houston, Texas, I was involved in product formulation and had the opportunity to study the remarkable benefits of composting and mulching on a unprecedented scale. Living Earth Technology is today one of the largest manufacturers of compost, mulch, soils, soil amendments and growing media in the nation, processing hundreds of thousands of cubic yards annually. Using a natural, organic mulch such as compost, aged bark or leaf mold provides many benefits. Here are some of the major contributions mulch can make when applied around your oleanders and other plants.

1. Retains moisture: A 3 to 4 inch layer of mulch will help retain soil moisture, especially in the critical fibrous root zone in the top 6 inches or so, will prevent root scorch from excessive surface radiation, and reduce stress resulting from rapidly alternating cycles of watering and drying out.

2. Moderates soil temperature: Mulch is of significant value in cooling the surface layers of the soil in summer and keeping the root zone warmer in winter. To make your own test of the amount of temperature difference mulch can provide, on a hot summer day place your hand on an exposed area of soil and then place it under a 3 to 4 inch layer of mulch, also in an exposed location. We have often registered a difference of more than 40 degrees!

3. Suppresses weeds: A thick layer of mulch does an excellent job of suppressing weeds, especially a material with interlocking fibers such as that made from hardwood. Any mulch, however, is useful since it blocks the sunlight and inhibits weed seeds from germinating.

4. Adds valuable organic matter to soils: As mulch breaks down it adds valuable organics to the surface area of the soil, much as leaf litter does in a forest. Adding mulch each year improves the tilth of the soil and aids in building populations of beneficial soil organisms.

5. Improves appearance: A well-mulched bed or island of shrubs helps beautify any landscape. The neat, clean appearance will last for months with minimal maintenance.

6. Protects against soil erosion: On slopes and banks mulch is an invaluable asset in protecting against soil erosion. Aside from the obvious benefits of this in itself, it permits plants to establish more rapidly and provides root protection for years to come.

Tree form oleander - elegant and refined.

Oleander flowers and follicles (seed pods) photographed on Grand Cayman.

Bouquet pruning - allows light around the base and enables one to see into the landscape.

Umbrella form

Pruning

In the following section we discuss general and specific types of pruning for oleanders. The most natural form of an oleander is globular, but large varieties allowed to grow as globular forms take up a vast amount of space and often obstruct the view. There are ordinances in cities throughout the country that limit the size of shrubs planted in medians and on corner lots in order to maintain visibility for motorists. Another negative in allowing some of the large oleanders to grow in their natural globular shape with dense foliage that extends to the ground is that they make excellent hiding places for thieves and muggers! Though dwarf forms are available and more are being developed through hybridization, we need not to be limited to these if our favorite flower happens to be on a larger variety since oleanders respond so well to various pruning treatments.

General Guidelines

Pruning requires considerable forethought to achieve the best aesthetic appearance for plants in your landscape. There are two basic types of pruning recommended for oleanders. The first is thinning by removing stems at ground level. Stems that are old, woody, or dead are always removed first in the thinning process. The second is "heading back" by cutting healthy stems to a point where you create a harmonious balance between size and shape. With annual pruning one can keep any oleander to approximately 60% of its ultimate height but we recommend never to remove more than 1/3 of its mass at a time. If you know the ultimate size of the plant you wish to have in a given space, choosing a cultivar with a similar natural habit will save much pruning labor in the future. We live by two pruning rules for most plants, but especially for oleanders. Never use hedge shears on oleanders! And remember that a well-pruned plant should appear as if it just grew that way naturally.

Clarence Pleasants remarked, "Of course, the real secret to the height of oleanders is pruning them now and then, not just a French haircut, but after they're a certain height, say six or eight feet (that's a general height), then it's wise to go and take some of the older stems out and encourage some of the younger ones to grow. That's the easy way to keep the size down, if you trim two or three stems out every year."

In Galveston and areas with similar climates oleanders should not be pruned in the winter or spring. With the exception of the everblooming varieties, we suggest pruning no later than the first week in September and as early as mid-July when many plants finish blooming. The reason for this pruning schedule is that in warm climates oleanders tend to put on a spurt of growth, both leaves and flower buds, for two or three weeks in the early fall after the summer's heat and with the onset of cooler days and nights. Also, with the coldest weather normally occurring in December and January, new growth will have time to harden off before winter. Plants that bloom only in spring may be pruned any time after flowering but early to midsummer is best to avoid removing the wood that will produce blooms the following season. In colder areas plants should be pruned as early as possible in spring, especially to remove any winter damage the plant has incurred. The key point to remember is that oleanders carry latent flower buds on the mature wood. Pruning in late summer, or early fall at the very latest, allows new growth to regenerate and harden off. Pruning in winter or spring removes the mature, flowering wood and delays bloom until new growth matures.

There are many other reasons for pruning in addition to restricting size. Oleanders that bloom abundantly seem to reinvigorate themselves and come back more readily after heavy pruning. Any time an oleander is cut or frozen to the ground it will grow back from the base with a myriad of small stems creating the globular shape mentioned above. To reduce the size of a plant at the base, basal suckers should be pulled off, not cut, to discourage regeneration. For oleanders that have compact forms, annual pruning is recommended to prevent legginess, especially if plants are not grown in full sun. When plants that have been grown outside most of the year are moved into a greenhouse for winter protection, they tend to stretch making pruning necessary. Ted Turner, Jr. has found it expedient to prune three times during the winter, allowing the plant to break each time between prunings to ensure a full plant the next year. One should always prune to remove weak, dead and crossing branches, diseased parts such as gall infestations, unhealthy and unsightly areas of freeze damage, or brown leaves due to wind and salt burn. Since a tremendous amount of a plant's energy is expended in the production of seeds, pruning to remove seed pods will increase the number of blooms per plant and extend the flowering season.

Dwarf and Petite Oleanders

An expert in the culture of dwarfs and petites, Ted Turner, Sr. recommends pruning dwarf varieties fairly severely in the spring after all danger of frost is past. "Of course, any time you prune a plant you stimulate growth to a degree," he comments. "With oleanders you definitely do. It stimulates the growth and they're going to sprout and then, if there is a late freeze, you get hurt. But you still need to prune them as soon as possible so that you'll get blossoms at least by mid-spring."

For a long time Ted Jr. didn't recommend pruning the petites such as 'Carnival' and 'Petite Salmon', primarily because they never stop flowering. If there was no winter freeze the buds that were on the plant in late fall would open in early spring and the plants would bloom continuously through the summer. But he noticed that the old inflorescence stalks remained and looked like "fuzzy candles" sticking out four or five inches on the plant. Now he recommends giving them a "haircut" in the spring if there has not been a freeze.

Shrub Forms

To maintain oleanders in their natural globular form, prune back individual branches a few at a time to enhance the globular appearance. Never shear the leaves and growing tips to achieve a rounded form because you will not only have a plant with unsightly clipped foliage but you will have also cut away most of the flower buds.

Multi-trunk Forms

Select any number (aesthetically it is best to choose an odd number) of the healthiest and most attractive stems to create a multi-trunk. Remove all the remaining stems and the lower side branches from the chosen stems so that the plants develop a full branching head at the height desired. Maintenance on this form is very high as new stems continually develop from the roots and new shoots emerge from the stems that were cut to the ground. If not regularly and assiduously pruned, an unsightly "window" effect is created. (See photo pg. 85) If, however, you begin with a young plant and pinch, rub out, or excise completely with a budding knife the latent or newly emerging leaf and stem buds, the task is made much easier.

Multi-trunk specimens are one of Ted Turner, Sr.'s specialties. "I love the multi-trunk oleanders, I think they are gorgeous," he says. "It takes a lot of work to keep the sprouts off them, but after they get real, real old they don't sprout anymore." Another reason for creating multi-trunk forms in areas where there are high winds is that three or more trunks buffer each other and withstand the wind much better.

Bouquet Forms

This form provides an alternative to the natural, massive, globular shape of the larger oleander varieties. It allows light and air around the base of the plant as well as a much more open view, creating a vase-like appearance rather than a solid screen. Pruning a bouquet form begins with the standard pruning practice of removing dead or dying, weak and crossing branches, then reducing the overall size if desired. After this, remove at ground level as many of the outer stems as necessary to create a narrow base. It may be preferable to achieve this by also pruning away some of the older stems in the center of the plant leaving younger stems that will grow to take their place. If you simply cut back all the longest stems to the desired height, your plant will look like cluster of a woody canes with unsightly tufts at the top. To enhance the "bouquet" appearance, head back some of the remaining outer stems. This will further reduce the plant's diameter and initiate a flush of new leaves. (See photo, pg. 80)

Umbrella Forms

This technique involves binding several stems together at the desired height creating the appearance of a large single trunk, somewhat like the swollen caudexes of many tropical trees. Plants are allowed to branch just above the point where they are bound and arch outward to create this distinctive and eye-catching form. (See photo, pg. 80)

Tree Forms, Patio Trees or Standards

The only tree forms available to the public for many years were the "patio trees" from Monrovia Nursery in Azusa, California. These were pruned with six feet of trunk. Ted Turner, Jr. remarked that plants pruned to this height would have to be permanently staked in Corpus Christi. All tree form oleanders must be staked unless they are in a totally protected area away from gusts and high winds that can easily snap off the head. Stake the trunks with a steel reinforcing rod or galvanized pipe driven down parallel to the trunk of the tree and fairly close to it. Secure the tree at the height of the first branches to keep the head from twisting. Keeping the heads pruned a bit close to prevent them from becoming too massive will lower wind resistance. In windy areas patio trees should have trunks no higher than 3 to 4 feet and should be planted in a protected spot. Oleanders trained in this manner take up much less space than globular or bouquet forms and are appropriate for gardens with limited space. Since they cast very little shade one can grow a large selection of low-growing groundcovers or colorful annuals beneath them. Oleander trees in bloom are a magnificent sight but one must follow the laws of Nature to create this unique form.

It is not possible to create a tree form by cutting away all the stems of a shrub except one. With an established oleander this would create a pruning nightmare worthy of a Chaplin film with new shoots growing back faster than one could remove them. The ideal way is to begin with a large, straight, rooted cutting or a young whip on which you can rub out any emerging new growth. This will channel the energy up the trunk and you can allow the plant to branch at whatever height you prefer. There are two techniques for achieving tree forms that we recommend for simplicity and dependability. The first method, utilized by Gary Outenreath at Moody Gardens, is to crowd together a number of oleander seedlings or cuttings. The plants will grow straight upwards, stretching to attain the sunlight, and being crowded together few side shoots will develop. Those that do can easily be rubbed out. Plants grown in this manner do not tend to sucker at the base. When they achieve the height you want, pinch out the tip and they will branch. This same method is used for hibiscus. The second technique is to layer (see sections on air layering and ground layering) a stem of the height and diameter you want. Since oleanders grow rapidly, especially in warm climates, you can literally watch your tree develop in one or two seasons. Remember, too, that you have a wide palette to choose from, not only in color selection but in the different sized cultivars that are available.

An innovative technique Bob Newding uses for growing a tree form is to slit a piece of PVC pipe along its length and fit a sleeve around the base of the stem. This accomplishes two useful things; it prevents damage to the tree from string trimmers and will shade the trunk, inhibiting shoots from emerging from the base and sides. (photo, pg. 74)

Excellent examples of tree form oleanders may be seen at the entrance to Moody Gardens in Galveston, Texas, and in many other landscaped parks, avenues and residences throughout the city. They are also used with stunning effect at Disneyland and Disneyworld. Tree forms provide color over a long growing season and are especially beautiful in flower beds and lawns. Since oleanders are evergreen they are not messy due to excessive leaf drop. The tree form, being less massive, is also a good choice in an area where one might wish to avoid obstructing a view. We have seen many elegant specimens planted in large con-

Young standards (tree forms) being trained at Turner's Gardenland, Corpus Christi.

Multi-stems at Galveston College.

Multi-stems in need of pruning to remove basal growth.

tainers in front of office buildings and in landscaped settings around malls in southern California.

Hedges and Screens

It has been Bob Newding's experience that if you plant varieties with different mature heights when creating a hedge or screen, the larger plants will take over unless all are pruned in accordance with each variety's ultimate height. By pruning and shaping each plant as an individual, you will create an interesting grouping with varying heights and colors. Planting only one variety simplifies pruning but creates the completely different look of a uniform hedge. When closely planted, oleander hedges eventually become virtually impenetrable, an effective screen against foot traffic, animals, and even dust and noise.

Root Pruning

Root pruning is another method of keeping plants to size. However, this is much easier said than done unless plants are fairly young. Using a spade, cut down vertically around the drip line of the plant to the depth of the spade. This will limit horizontal root growth and thereby keep plant size restrained.

Hybridizing

As a result of discussions with a friend at Longwood Gardens in Pennsylvania, Clarence Pleasants was encouraged to begin hybridizing with an aim toward first developing greater hardiness, extending the growing range of oleanders into cooler zones, and then to seek good flowering qualities. In hybridizing, one must begin with a clear goal and select existing plants that have the inherent genetic qualities to lead to the realization of that goal. As an example, Ted Turner, Sr. wanted to develop a free-blooming plant with red flowers that was more dwarfed and compact than any existing cultivars. Thus far he has succeeded in two of his three goals and is still working towards achieving the elusive red color. Monrovia Nursery has concentrated on free-blooming plants that are compact in size and have unique, almost fluorescent colors.

Most of the cultivars we have listed are the result of pollination by insects, namely butterflies and moths, as self-pollination of oleanders rarely occurs in Nature. In more than a hundred hand-pollinations, Frank Pagen found that all cross-pollinations between cultivars appeared to be compatible. "In every combination of cultivars tested, the pollen tubes reached the ovules without difficulty," he reported. This section is based, in large part, on the studies of Pagen and the following is quoted with his kind permission. It is included to inspire those who would like to develop new hybrids by describing the procedure for successful pollination. (Please refer to the excellent line drawings by Pagen, pg. 87, 88)

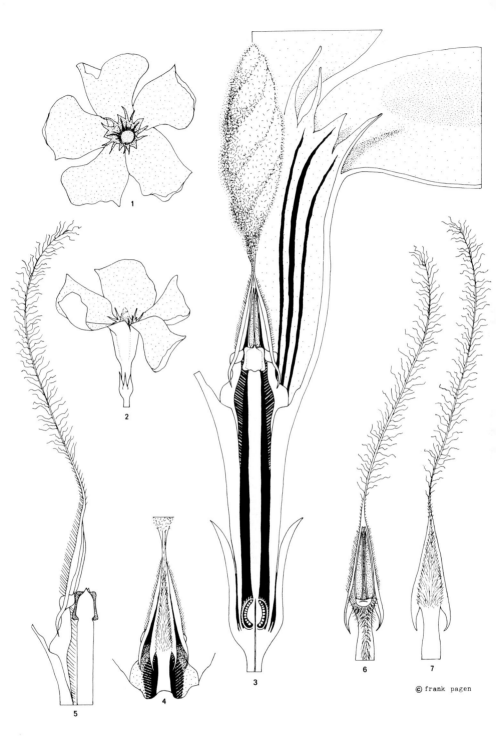

The flower of Nerium oleander L. 1. corolla from above x ¾; 2. lateral view x ¾; 3. lengthwise section of the flower tube, 2 anthers removed x 6; 4. cone of anthers x 7; 5. position of anther and pistil head x 7; 6. anther, adaxial view x 7; 7. id., abaxial view x 7 (drawn from living plants).

Illustration courtesy F.J.J. Pagen

87

H. van Rheede tot Drakestein. 1689. Hortus Malabaricus, Vol. 9. t. l: "Tsjovánna-arelí".
Copper engraving.

Illustration courtesy F.J.J. Pagen

The flower of *Nerium oleander* features a pistil with a morphologically highly differentiated pistil head.... The entire pistil head is covered with bristly hairs, which determine, by their length, the specific shape of the pistil head. The central cylindrical part of the pistil head is covered with rather short hairs, inserted on a glandular epithelium which produces an adhesive. At the upper edge of the cylindrical part the hairs are longer, pointing outward and upward, thus forming a ring or wreath. On the lower part of the pistil head long hairs, pointing downward, are arranged in a sinuous line around the cylinder, forming a collar. Below this collar, which hangs down from the pistil head, is a circular depression marking the transition to the style. The pistil head is topped by two short erect appendages.

This morphological differentiation is accompanied by an analogous functional differentiation. Only specific areas on the pistil head have a stigmatic function and are receptive for pollen tubes.

. . . .

The receptive stigmatic area evidently is located on the circular depression, right below the collar on the lower part of the pistil head....
Pollen grains will germinate on other parts of the moist surface of the pistil head as well, but the pollen tubes will grow along this surface without penetrating it. Penetration is exclusively possible in the indicated area.

. . . .

The *Nerium oleander* flower has an intricate construction connected with a refined pollination mechanism.

Dark, usually carmine red, longitudinal lines in the corolla throat point to the nectaries around the ovary at the base of the funnel-shaped corolla tube. However, much of the passage is blocked by the corona lobes and by the woolly plug formed by the intertwisted appendages of the anthers. The short-stalked stamens are inserted at the middle of the corolla tube. The anthers have firm sagittate plates outside and are introrse, adhering to the pistil head, thus forming a cone in the center of the flower. The pollen is shed inside the cone of anthers and collects on the top of the pistil head, completely isolated from the stigmatic area below the collar at the lower part of the pistil head.

Between the five triangular anthers are chinks, narrowing to the top of the cone. The insect's tongue can be inserted in the tube through these chinks or through the openings between the filaments. In either case, when the tongue is withdrawn it gets jammed between the adjacent anthers. A powerful pull is necessary to release the tongue and in this action the tongue moves along the cylindrical part of the pistil head, where it gets covered with adhesive, and then through the pollen chamber, where pollen grains are glued to the tongue. The insect is probably disturbed and leaves the flower. In another oleander flower the procedure is repeated and when the tongue is withdrawn it brushes past the collar hanging down from the pistil head. Thus, the pollen is deposited below the collar, exactly on the receptive stigmatic area. Moving upward the tongue passes again along the cylindrical part of the pistil head and through the pollen chamber. The tongue is covered again with pollen and the insect is ready to pollinate another oleander flower.

The *Nerium oleander* flower is adapted to pollination by Lepidoptera (butterflies and moths). Only these insects have a tongue long enough to reach the bottom of the corolla tube and have the strength and endurance need to release their tongue from the trap and establish pollination of the oleander flower.

Pagen utilized fluorescent microscopy to locate the stigmatic area on the pistil of oleanders. Through the experiments he conducted, the exact location of the receptive area was pinpointed as was the compatibility of self-pollinations and cross-pollinations between cultivars. Fourteen single-flowered plants provided by the Dutch Oleander Society were the basis for the experiment.

Ted Turner, Sr. often jokes with people that he uses "Q-tips" to pollinate oleanders, but he becomes excited and almost poetic when he describes a cross section of a flower under a microscope. "I think that is the most fascinating thing in the world — it looks like a pot of gold on top of the stamens!"

Now that we have a clear understanding of how to pollinate an oleander, we can proceed by selecting two newly opened flowers. We have not studied the period of receptivity of oleanders, but have found with plumerias, hibiscus and many other tropical plants that the highest receptivity is around 10:00 a.m., so this would probably be a good time to begin. Be sure to emasculate the flowers at the beginning to prevent any possibility of self-pollination and cover them when finished with fine gauze or cheesecloth to prevent any other pollen from entering. Tag the pollinated flowers, indicating first the seed parent followed by the pollen parent and the date. Example: 'Petite Pink' (seed) x 'Sorrento' (pollen).

A Perfect Lady

I knew a girl who was so pure
She couldn't say the word manure.
Indeed, her modesty was such
She wouldn't pass a rabbit-hutch;
And butterflies upon the wing
Would make her blush like anything.

That lady is a gardener now,
And all her views have changed, somehow.
She squashes greenfly with her thumb,
And knows how little snowdrops come:
In fact, the garden she has got,
Has broadened out her mind a lot.

Reginald Arkell

Dwarfing Plants with Growth Retardants

Oleanders may be dwarfed by using the commercially available growth retardant, Cyclocel. From rooted tip cuttings to finished plants takes about one year and requires at least three applications. Considerable experience is necessary to learn when to apply as well as the proper concentration. If the retardant is not continually applied at regular intervals plants will resume their normal growth rates, so for most oleander enthusiasts, choosing a dwarf or compact variety is a more practical alternative.

Pests, Diseases, Problems and Solutions

There is virtually no insect or disease that can kill an oleander with the exception of root rot diseases caused by poor drainage and a rare canker that attacks severely stressed plants. In fact, chemicals used to treat insect attacks may occasionally cause more damage than the insect itself. The Turners caution against using Diazinon on oleanders as this chemical, in their experience, will kill them, even if spray drifts over when spraying other plants. Although oleanders are noted for their resistance to pests and diseases, they do suffer occasional attacks. We encountered a twig girdler in India that would cut around oleander stems and decimate plants quite rapidly but new growth would simply sprout below the girdled area. In general, insects favor other plants and will attack them first. The Turners note that if you have mites on an oleander, all you have to do is move a *Datura* (Angels Trumpet) or *Chamadorea* (Parlor Palm) next to it and they'll leave the oleander in a hurry. Ted says that you can almost see them jumping down! Most pests are fairly easy to control and although some problems are more serious than others it should be remembered that, by and large, regardless of the rather extensive list that follows, there are few plants that give so much beauty with so little care.

Aphids: Aphids are probably the most common insect pest of oleanders. These sucking insects feed on the sap of leaves and flower buds and cause them to be deformed. There is even an aphid named for oleanders, the yellow-bodied oleander aphid *Aphis nerii.* In a most interesting article in *Nerium News,* (Spring-summer '93), Ken Steblein writes of the Integrated Pest Management (IPM) program at Disneyworld. In addition to beautifying the landscape, oleanders are used as a host plant for the oleander aphid which feeds exclusively on oleander leaves. The IPM department introduces to the oleander plants the tiny parasitic wasp *Aphidus matricatriae* which lays eggs on the aphid for approximately two weeks and causes the aphid's body to be covered with a hard shell like a "mummy." At temperatures around seventy degrees, the eggs hatch inside the aphid shell and feed for about two more weeks. Oleander leaves with aphid "mummies" are then placed among the vegetable crops in the Epcot Center's Economic Greenhouse. The wasps mature, eat a hole in the aphid "mummy" and seek out aphids on the vegetables where they repeat the cycle. Oleanders are used again as hosts when necessary to maintain a constant supply of the parasitic wasps.

On a smaller scale, aphids are fairly easy to control in the early stages of infestation by simply washing them off with a strong stream of water. If you use a sprayer, add a teaspoon of liquid dishwashing detergent per gallon of water and spray with maximum pressure. If you experience a heavier infestation, you can get things under control with one of the following formulas:

1. Two tablespoons each plain vinegar and household ammonia per gallon of water. For more effective application use a surfactant such as liquid dishwashing detergent or a commercially available spreader-sticker which, in addition to breaking the surface tension of the water (as detergent does) to provide better coverage, will also adhere the material to the leaf and stem surfaces.

2. If you have a serious infestation, Malathion (applied exactly according to directions) has proven effective, especially in combination with a surfactant.

3. The Turners use Phycam, a wettable powder, to control aphids, spraying very early in the morning when it's cool. They have found that powdered insecticides tend to be less injurious to plants than oil based sprays and caution that the use of Sevin on petite oleanders can cause the growing tips to burn.

Ants are known to transport, or "farm out," insects such as aphids to different areas of a plant and, in return, feed on their honeydew secretions. It is therefore important to look for ants when you see the first sign of aphids and to treat them as well.

Canker: In the spring of 1989 there were isolated reports from the Texas gulf coast, including the Galveston Country Club, of a mysterious disease affecting established oleanders. According to an article in *Nerium News* (Summer '89), in the instances observed, leaves turned gray and then yellow before falling to the ground and eventually the entire plant died rather suddenly. The problem was identified by the Plant Diagnostic Laboratory at Texas A&M University as Botryodiplodia, a weak, opportunistic fungus or canker that damages the vascular system of the plant. The fungus is ubiquitous, but in order to enter a plant it must be accompanied by stress factors such as freeze, drought, high wind, nutrient deficiency, chemical damage, etc. Though described as a weak fungus, it does not readily respond to available fungicides. Some control was achieved by spraying with a fungicide containing Benomyl every two weeks and also with Kocide 101, but it was equally important to meticulously remove any and all symptomatic branches. It was observed that the worst infections occurred on pink and red varieties. Always handle and destroy infected prunings with care to avoid contaminating other plants, but do not burn because of the toxic fumes from the oleander wood. Disinfect pruning tools with a 10% solution of chlorine bleach.

Caterpillars: To date, four species of caterpillars have been observed feeding on oleanders. One, the Oleander Caterpillar *Syntomeida epilais jucundissima*, is a stinging type that should be avoided. It is a serious problem in Florida and can rapidly defoliate plants if not treated, but can usually be checked with one application of Sevin or Malathion. We have no reports of it being a problem in Galveston or other areas of the United States. Our recommendation is to spray with *Bacillus thuringiensis* which is harmless to humans but highly effective against all kinds of caterpillars.

Dodder: Dodder is a climbing parasitic plant that has been observed on oleanders growing on the causeway entering Galveston. In appearance it resembles a tangled clump of yellow-orange strings. The parasite reproduces itself by seed and although it does not seem to spread rapidly enough to become a serious threat, it should be removed and destroyed.

Fungal Diseases: Powdery mildew, leaf spot (black spot), and root rot can all be prevented or severely curtailed by keeping containerized plants on the dry side during the winter and by providing adequate drainage to plants grown in the ground. Kocide 101 is a good all-purpose, copper-based fungicide that is effective and fairly safe to use.

In a nursery situation where many pots are close together in an area, the Turners have often had problems with two types of fungus that attack the petite oleanders, especially those shipped in from the west coast. One appears similar to dampening off fungus and the other attacks the stems, quickly spreading from one plant to another. They have observed this on 10% to 15% of the gallon size material coming from California, but when the plants are planted out in the landscape the

Petite Pink

problem does not occur, seeming to be specific to nursery conditions. They have found Ornalin to be an effective control but advise never to use Benomyl on oleanders for control of fungus. Once, when Ted Sr. was spraying Benomyl on other plants in his greenhouse, he accidentally sprayed some oleanders and lost 14,000 cuttings at one time!

Mealybugs: These are small, soft-bodied, sucking insects covered with a white, cotton-like substance. Two species have been reported on oleanders. For effective control use the same formulas as mentioned above for aphids.

Nematodes: Clarence Pleasants noted that nematodes were found in Galveston but this is the only report to date and the incidence appears to be very localized.

Oleander gall: The bane of oleanders is a wart-like, bacterial gall (*Pseudomonas syringae*) that appears on succulent, non-woody parts of the plant such as leaves and younger stems. Although not lethal, it is very unsightly causing streaks of swollen tissue or numerous knots along the infected area and can even cause new leaves to deform. It is easily spread, especially by infected pruning tools, and there is no chemical cure for it. In an article appearing in the *Galveston Daily News*, November 20, 1991, Dr. William M. Johnson, a plant pathologist with the Galveston County Extension Office, concurred that the gall is caused by a bacterium and explained that it first multiplies within the gall tissue, then migrates to the surface where it is easily spread to other plants or parts of the same plant. Gary Outenreath informs us that it is both airborne and carried by insects and that it enters through the flowers, distorting and often blackening them. The Turners find that in Corpus Christi it is more prevalent during a cool spring with very cloudy days and temperatures around 40 degrees, and is occasionally seen under foggy conditions on plants near water. If a plant is infested with gall, the infected parts must be carefully removed and completely destroyed so as not to spread the bacteria. Pruning tools should be dipped in a 10% solution of chlorine bleach each time a cut is made. This may be laborious but extreme care must be taken with a bacteria that is so highly contagious. If a plant is heavily infested it may be necessary to cut it as close to the ground as possible and remove all the leaf litter as bacterial spores are also in the litter. Spray the exposed stump with a 10% solution of chlorine bleach. In one growing season the plant can put on four to five feet of beautiful new globular growth and in two years will reach its full height once again. The best time to look for gall is mid-summer after many plants have finished their major flush of bloom and after especially rainy or humid weather.

Root Rot: This is a common cultural problem usually caused by overwatering plants in containers or providing insufficient drainage for plants in the landscape.

Scale insects: According to the International Oleander Society, seven different species of scale insects have been identified on oleanders. They are soft-bodied, sucking insects that subsist on the juices in plants and secrete a waxy substance over themselves that affords them protection. Scale insects are often very difficult to control and can do serious damage if not treated. In order to effectively destroy scale populations, their protective coating must be penetrated. The spray formulations mentioned above for aphids should do the job but be especially careful to thoroughly spray the backs of leaves since this is where scale is normally attached. Scale insects can weaken a plant severely so that it will be susceptible to a host of other insects and/or diseases. The Turners have used oil sprays successfully but they offer some cautionary advice. Ted Sr. says, "Leave it on about an hour, then wash it off if you don't want to hurt anything. That's long enough to kill any kind of insect that's on there. I never was one for leaving insecticide on plants all day long. I hate to spray for anything but I have to now and then."

Snails: Snails and slugs are usually only a problem in spring as they will feed on flowers but can be controlled naturally by sinking saucers of beer in the ground, or chemically with a variety of commercially available baits placed around the plant.

Spider Mites: Spider mites are rarely encountered on plants growing outdoors but can be a problem in enclosed areas such as greenhouses or wherever ventilation is inadequate. Populations can increase rapidly, especially during hot weather, and the mites are so tiny they often go unobserved until a serious infestation is underway. Kelthane is effective in the control of spider mites in greenhouse conditions but it is best to follow the Turners' advice and apply all chemicals when temperatures are cool and then wash them off after an hour or two.

Witches broom: Witches broom is an abnormal growth of a number of shoots that may grow in a similar direction or may become a tangled mass, hence the name. It has numerous possible causes, the primary ones being infection by various fungi such as rust, mite infestations in the growing tip of the plant, parasites and, on occasion, low temperatures. The form of witches broom that attacks oleanders is unsightly but not fatal, and can be controlled by simply cutting out and destroying the affected parts.

Yellowing of leaves: Chlorosis, which is a sign of iron deficiency indicated by yellow leaves with green veins, and yellowing leaves due to lack of nitrogen are not usually a problem with oleanders. If you should observe chlorosis, apply a chelated trace element mix either in granular or, for faster results, water-soluble form. Nitrogen deficiency is easily remedied by an application of fertilizer as mentioned previously. Occasionally the lower leaves on plants will turn yellow during periods of drought or intense heat. This is a natural phenomena, the plant's way of dealing with stress. With the onset of cooler weather and/or sufficient moisture, the yellow leaves will fall to the ground or can be washed off with a hose as they are replaced by new growth.

Cold Tolerance, Hardiness

Of all the oleanders, those with red flowers appear to be the most cold tolerant. One indication of hardiness is the thickness of the leaves. The thin, light green leaves often found in the free-blooming varieties are generally indicative of more tender varieties and dark green, leathery leaves characterize hardier forms. The tender varieties are also more susceptible to wind burn. Clarence Pleasants offers

John Samuels

some fascinating insights on the ability of oleanders to become progressively hardier. "They have a tendency to adapt," he says. "Our [garden] supervisor, the man who encouraged me, would have us root these plants in spring and we'd grow them on. We didn't keep them inside long. After they were out of the 4 inch pot area and of a size ready to transplant, we'd plant them in a row in a raised bed for drainage out in the field, out where it's cold, cold, cold, in full sun and with frost to keep them from being too aggressive in growth. This was the experimental stage where he [his supervisor] felt you could find the hardiest and the best. Most of the plants would live and they'd grow very slow." Pleasants goes on to say that his supervisor considered the first couple of years to be the most important. All of the plants would die to the ground in the first winter and many would come back the following spring. In the second year many plants would not die completely to the ground and would have hardy growth alive throughout the winter. In the third year the remaining plants would be dug up and planted in a sunny area with poor soil, little fertilizer and little water. This treatment toughened the plants and led to many more hardy varieties. Some other salient observations Pleasants makes is that double-flowered varieties do not appear to be as hardy as single-flowered forms, and fragrant flowers seem to come from more tender parents or from a milder area of the world. This concurs with our experiences with oleanders in India and southeast Asia.

Certain oleander cultivars are able to withstand temperatures well below freezing. The RHS *Dictionary of Gardening* mentions oleanders tolerating 14° F for short periods. As suggested above, hardiness among cultivars is highly variable. Some tender varieties are suitable only for growing in containers or the warmest zones while others show no sign of damage during lengthy periods of cold such as the December 1983 freeze in Texas.

Cold tolerance can be increased by withholding water and fertilizer in late summer and early fall, allowing new growth to harden off before the onset of cold weather. It has been observed in Galveston that when many varieties are subjected to cold they are stimulated the next spring to grow and flower even more abundantly than before.

Drought Tolerance

As with plumerias and many other members of the Apocynaceae family, oleanders are remarkably drought tolerant. The Turners relate an experience they once had trying to destroy a surfeit of undesirable plants: "You can't kill them with Round_p like you think you can. We'd dump them out on a road to let them expire and we'd come back and they were still alive. They're extremely drought tolerant. You can't kill them once they get started!" The Turners also note that even if the leaves of plants are burned from drought, it is still possible to take cuttings from them.

Salt Tolerance

Although we have read some accounts from the Caribbean Islands stating that oleanders cannot grow in saline areas, this is contrary to the Turners' experience in Corpus Christi and to Newding's in Galveston. The Turners have found that not only are the leaves salt tolerant, the roots are as well. In many areas of Corpus Christi one can literally see the salt buildup as a crust on top of the soil. Ted Sr. notes that oleanders are one of the few plants that can survive under these conditions. (See also Chapter 9, "Oleanders in the Landscape".)

Overwintering and Freezes

Oleanders tend to become dormant with the onset of the first cold weather and should not be watered or fertilized until new growth begins in spring. The Turners have saved thousands of hibiscus, bougainvilleas and oleanders that have frozen to the ground with a technique they discovered. After the freeze, rake the soil back exposing some underground stems, then wash the soil away. New growth will be stimulated to emerge from the tops of the anchor roots below the damaged area. The tender new growth that regenerates will be very susceptible to any subsequent freezes and should be mulched heavily, at least ten to twelve inches, if freezing temperatures recur. Be sure to remove the mulch as soon as temperatures permit, however, to again expose the roots to the sun.

Rose collars about 6 inches in length can be placed around oleander stems at ground level and filled with mulch or soil to keep the base of the plant from freezing. Protecting this much of the stems of small varieties enables them to produce new growth and even bloom again within a few months. Larger varieties may take longer but will still recover much sooner than if frozen to the ground. Another alternative for a plant that is too large to cover is to cut it down to a size you can cover! — in a hard freeze the top growth would probably be lost anyway. A common winter treatment for roses is also effective; simply cover as much of the base of the plant as possible with soil and remove when freezing weather is past.

In summary, oleanders can endure several degrees of frost and even repeated freezes as long as they are brief. Problems occur when low temperatures last more than two or three days.

Growing Oleanders in Containers

Americans have been growing oleanders in containers since they were brought across the sea by early settlers from the Mediterranean, but their history as cherished container plants in Europe and the British Isles stretches back much farther than that. Placed in formal planters on steps and terraces in full sun out of doors in summer, and mixed with flowering tropicals and subtropicals such as hibiscus and bougainvilleas in a glass house or conservatory during winter, oleanders have long been considered one of the finest plants for container culture.

We have mentioned that tree forms are excellent for pots and planters as are the more compact varieties, but any oleander

Turner's Kim Bell™

can be grown in a container with occasional pruning and shaping. As corresponding secretary for the International Oleander Society, Elizabeth Head has received letters from people throughout the United States telling of their successes with pot culture. One is from a gardener in Michigan who has oleanders in full bloom all summer in tubs along his driveway. The petites and dwarf forms are especially suited for container culture, not only because they require less pruning, but because they will often flower prolifically even in small pots. In a warm climate one can transplant an oleander liner (2¼ inch pot) into a one gallon container on March 1st and it will be blooming in April. Also, because they don't require large containers, they are easily moved around. Ted Turner enthusiastically comments, "You should see the 'Petite Salmon', the original that came from the [Los Angeles State and County] Arboretum! If you'll put three gallon-can plants in a 14 inch container, you've got an oleander that is over 3 feet across and 2½ feet tall and you just can't beat that out on the decks, out over the salt water where the decks go out over the canals. You can't beat them out there — they'll take the wind, they'll take the sun, and they'll take the salt breeze much better than al-most anything else."

A plumeria enthusiast for many years, our friend Jim Nicholas subsequently became interested in oleanders and has shared some fascinating experiences grow-ing them in 6 to 8 inch containers in Connecticut. Jim usually uses sand as a rooting medium but told us about once receiving a package of cuttings only to find he had no sand, so he put them in water. After just four days he discovered that 'Ed Barr' was already putting out roots! He has found that the easiest and quickest varieties to root, even during the winter, are 'Little Red', 'Ed Barr' and 'Magnolia Willis Sealy'. After rooting, Jim moves the plants into small containers using a very light potting soil or a soilless mix to which he adds sand for added drainage and stability. He has successfully flowered all the varieties he has tried; some, such as 'Magnolia Willis Sealy', even bloom profusely in the winter under Gro-lights. But the secret to flow-ering plants in small pots, he says, is in the fertilizer. By applying a formula low in nitrogen to restrict stem growth and high in phosphorus to encourage bloom, Jim has plants no larger than 2½ feet tall flowering prolifically nearly year-round. Smaller plants are overwintered under Gro-lights where they continue to grow and often bloom, others are kept in a west window where they rest without shedding leaves until spring. The only insect problems Jim has experienced have been one incidence of scale on a plant already weakened with a physical injury to the bark, and a few aphids which he says are a feast for the ladybugs that appear instantly when he puts the plants outdoors in the spring.

Choosing a Container

As mentioned earlier, oleanders develop many fibrous roots that spread later-ally but, in general, do not tend to grow very deep. For this reason it is best to plant them in containers that are wider than they are tall. Azalea pots are excellent choices as are tubs and whiskey barrels.

Soil Mixes for Containers

Although it may be necessary to water more frequently in the hottest weather, we recommend a highly porous, well-drained mix for oleanders grown in containers. A successful mix should contain ample amounts of compost or well-decomposed organic matter, sharp sand, a time-release fertilizer and fine aged bark. A small amount of sterilized soil can be added to a mix but it is best not to include local topsoil as it may contain bacteria, fungi or nematodes. More plants are killed by overwatering (too much TLC) than any other cause so bark nuggets or clay crockery pieces should be placed at the bottom of the container for added drainage. Always mulch the top of the container with a 2 to 3 inch layer of compost or well-decomposed organic matter to help retain moisture and to protect feeder roots from excessive heat.

Repotting Container Grown Oleanders

Horticultural rules often change over the decades, sometimes reverting to practices of an earlier time, sometimes advancing with newly acquired knowledge. Occasionally, two schools of thought exist concurrently. One concept in vogue today is to disturb root systems as little as possible and not to slice or comb encircling roots to avoid causing additional stress. Our practice has always been to seek to understand the nature of the plant and then, using care, take whatever steps are necessary to enhance its health, growth, flowering, fruiting, etc. We find that with oleanders, as with many other plants, the best practice is to comb down carefully without tearing and loosen all encircling roots, cutting off ones that cannot be redirected. After experimenting with many implements — knives, forks, three-pronged cultivators, weeding forks and more — we find that a short screwdriver (about 6" long) is the most effective and does the least damage. After loosening encircling roots, follow the same procedure as for transplanting. Be sure to have everything in readiness before you begin work. (See also the section on "Planting in the Ground".)

Oleanders don't mind being rootbound for a season or two, so transplanting is only needed when they outgrow their container creating an aesthetic imbalance, or when plant performance, even with supplemental fertilizing, indicates the need for a larger container.

Topdressing

If roots are pushing through the bottom of the container, repotting to the next larger size is advisable. If this is impractical, then topdressing by removing two or three inches of the old soil, including some roots, and adding compost or well-rotted manure will reinvigorate plants and suffice for another season. Topdressing should be done before the flower buds have formed. A second technique is to remove the root ball and cut back encircling roots or roots pushing through the drain holes in the container. Carefully remove a little soil from the base, top and sides of the rootball and replant with new soil.

Watering in Containers

During warm weather when plants are actively growing, water as often as necessary to supply roots with sufficient moisture and always water thoroughly. Allow the soil to become almost dry between waterings. Oleanders do not like "wet feet" so keep them on the dry side during cool and wet weather. Reduce watering significantly in early to mid-autumn, allowing new growth to harden off before winter.

Fertilizing in Containers

European growers recommend using a water-soluble fertilizer every two weeks once growth has begun in the spring and continuing this program through the end of summer. With the advances in timed-release or sustained-release fertilizers in the

100

1980's and 90's, we recommend alternating these with water-soluble formulas, especially those high in phosphorus since they are immediately available to the plant's system and insure a constant nutrient supply. For those wishing to avoid the use of chemicals, container culture is an ideal opportunity to try an organic fertilizing method we used successfully in India, "manure tea." The tea is made by soaking one part manure in three parts water for several days to a week. After soaking strain off the liquid, dilute by half if plants are very young and apply to the surface of the container. Each time you water, small amounts of nutrients will be dissolved and carried to the roots. Manure teas may also be made from other organic ingredients such as cottonseed meal and alfalfa pellets.

Pruning in Containers

There is a fair amount of contradictory advice on the subject of pruning plants in containers. For the most part, a common sense approach is all that's needed in addition to an understanding of the plant's nature. Pruning in late summer or early autumn will allow new growth to harden off before the onset of cold weather. Pruning in early spring to remove weak stems and old wood, and to achieve a desired height is also advisable. Periodic thinning will assure compactness and greater density. A helpful tip to remember is to remove any side shoots that develop just below the flower clusters in order to prevent flower buds from falling off. This is especially important for plants grown indoors.

Overwintering in Containers

Overwintering container grown plants is not difficult. A cool place such as a cellar, where temperatures may vary from the mid 30's to the low 50's F., is ideal. Although there is some disagreement on the amount of water to provide during winter dormancy, most growers advise that it should be little to none. As our experience with plumerias has shown, keeping plants on the dry side is the best course to follow. Plants stored in cellars during the winter should have some amount of light as well as adequate ventilation. Never fertilize during dormancy.

Sometimes an illuminating experience contradicts conventional wisdom and practice. One such case occurred during the hard freeze that hit southern Texas in December, 1983. We were living in Houston at the time and watched in disbelief as temperatures dropped into the single digits. In fact, in the Zone 9 area, the temperature did not get above freezing for five days and thousands of plants were killed. Farther south in Corpus Christi the Turners experienced similar temperatures, recording seven degrees at one point in front of their nursery. Ted Jr. commented, "Now, as far as I know, every oleander in this town was killed to the ground. However, I had a 'Petite Pink' in a container on the dock at my house and I can't remember the last time I watered it. I mean, it was so dry that it was crinkled! When I saw it still crinkled, I thought it was from the freeze, but when I watered it, it leafed right out to the ends of the branches." Ted Sr. remembers a few other petite oleanders that survived the freeze. One was 'Carnival' which he had planted in a container on the north side of a customer's residence. These were fairly young plants with a total height, including the pot, of not more than 2½ feet. The family went away on vacation for three or four weeks, returning after the freeze. During this time the plants received no water and were bone dry. A month or so later Ted visited the house and remarked that the oleanders in the pots had really grown back fast. The owner told him that they had never frozen! The Turners say that now they only water oleanders in the nursery when they expect a light freeze and heavy winds in order to prevent desiccation of the leaf surface; in a hard freeze they don't water at all. It should be remembered that this practice is only for oleanders in containers. Plants in the ground must be kept well-watered and heavily mulched before a freeze.

With the first sign of growth in spring, situate plants so they can receive the maximum amount of sunlight. A good general rule of thumb is to move oleanders outdoors when it is safe to plant tomatoes in the garden.

Tips for Year-round Container Culture Indoors

Plants grown permanently in greenhouses and garden rooms will necessarily have somewhat different requirements than those grown out-of-doors most of the

year but, in general, the basic principles regarding culture remain the same and many gardeners have great success growing oleanders indoors.

Experienced growers agree that most oleanders will bloom in a 6 to 8 inch pot with good care, but a 12 inch container is ideal if you want plants to attain greater mass. Dwarf forms are especially desirable for growing indoors year-round for the reasons described above. Soil requirements are essentially the same for all container grown oleanders in that the medium needs to be fertile, porous and very well-drained to be successful. Fertilizing should also follow the guidelines discussed above, bearing in

Turner's Daisy™

mind that growth rates may be somewhat slower during spring and summer and winter dormancy less pronounced.

Providing adequate light for plants such as oleanders that normally require a lot of sun is a challenge for many indoor gardeners. In our research we have found several references indicating that oleanders do very well under artificial lights. There are numerous books available on the subject of light gardening and equipment catalogs offer an assortment of fixtures, bulbs, light units, etc. We have not grown oleanders under artificial light, but for other plants we have found ordinary florescent fixtures with Gro-bulbs and a timer to be the simplest and most flexible arrangement.

Plant Selection Criteria

When purchasing oleanders at local nurseries take time to examine them carefully. Do not choose potbound plants where roots may be pushing through the drain holes or even at times visibly encircling themselves at the top of the container. Look for insects on stems and above and below leaves. Do not purchase plants with damaged or broken branches, die-back, symptoms of chlorosis or evidence of bacterial gall. Always select bushy, healthy plants with lush growth and refrain from "bargain" plants that may be weak, spindly or off-color. It will usually cost more in the long run to nurse such plants back to health. Choose the best plant for the intended planting site and avoid unreliable names such as Common Red, Hardy Pink, Double Yellow, etc. ·

Die when I may, I want it said of me
by those who know me best,
that I always plucked a thistle and planted a flower
where I thought a flower would grow.
Abraham Lincoln

A white wall, blue sky
- and the breathtaking beauty
of an oleander in full bloom.

Sue Hawley Oakes (cream yellow),
Pleasants Postoffice Pink (above),
Petite Pink (bottom right).

George Sealy - note Shasta daisies in foreground.

9

*And since to look at things in bloom
Fifty springs are little room,
About the woodlands I will go
To see the cherry hung with snow.*

*A.E. Housman,
A Shropshire Lad*

Oleanders in the Landscape

In a letter to Clarence Pleasants in Norfolk, Virginia, in July of 1964, Donald J. Moore, Superintendent of the Botanical Gardens in Bermuda wrote: "Bermuda is indeed famous for its oleanders. The plant is so prolific and well established that visitors may be forgiven for considering it a native, when in actual fact, it is an intro-duced subject. . . . Nathaniel Lord Britton, Ph.D, Sc.D., LL.D., one time Director-In-Chief of the New York Botanical Gardens, had this to say in his *Flora of Bermuda*, published in 1918 by Charles Scribner's Sons, New York: 'In nearly all situations except saline ones. Naturalized. Native of the Orient. Recorded as introduced to Bermuda in 1790, now one of its most beautiful floral features, blooming more or less throughout the year, most freely in Spring and Summer.'" Mr. Moore went on to mention that plants grow rapidly in Bermuda, are evergreen and found in single, semi-double and double forms in a wide range of colors, the most common being a pure pink and a pure white though the colour range extends through cream to deep red. Landscape uses of oleanders in Bermuda include hedges, individual specimens and colorful borders.

The oleander was perhaps not so loved in some areas of the United States in earlier times as we learn from this entreaty from the Royal Palm Nurseries in Florida, circa early 1900's:

> So many people, we have found, objected to this subject because they have only been accustomed to seeing great, scraggly, big-caned, ugly specimens in deserted hedge-rows or odd corners, the miserable victims of unpardonable neglect. As a matter of fact, this is one of the very love-liest of all flowering shrubs for general Florida planting and along the Gulf Coast generally, and the fact that it will grow in almost any soil and under almost any condition in the state makes it more than doubly valu-able. It is true that it will not be uninjured by cold in the more northerly sections, in severe freezes, but there is considerable difference in the rela-tive hardiness of the different sorts and, even when killed down occa-sionally, it comes right back under good culture and flowers freely as

before. The following varieties are more hardy than the others: Carneum, Frederick Guibert, Dr. Golfin, Savort, Single White, Dr. Brun.

While not suited for sheared hedge-work, they make splendid informal hedges — a double hedge of white Oleander background with single scarlet hibiscus foreground, or Carneum Oleander with pink Hibiscus foreground, and the like, make wonderfully effective enclosures for formal gardens, as screens for fences, and many other useful purposes. The flowers come in a wide range of color and in great profusion during the spring months, and some sorts bloom more or less throughout the summer. All shades make a splendid effect with the gray-green foliage which, even when there is no bloom, is attractive on properly kept specimens. The great trouble in the case of the Oleander — and indeed with most of our tropical shrubbery — is that the owner does not use the pruning shears! People who in the North would not think of neglecting their shrubs seem to take it for granted that under tropical conditions plants should grow and thrive and look well all the year round and under all conditions without any care. As a matter of fact they should receive as much attention here as elsewhere and under some conditions even more.

As the paragraph above suggests, the oleander is highly adaptable and tolerant of the most difficult conditions, including brackish water and heavy air pollution that would annihilate lesser plants. Epithets such as durable, tough, enduring and indispensable juxtaposed with terms such as splendid, long-flowering and handsome are commonly found in texts that mention the oleander as a mainstay in warm climate gardens throughout the world. Whether utilized as hedges or individually as focal point shrubs, pruned as single or multi-trunk standards, oleanders stand out against a backdrop of sea or sky, quickly cover fences, hide unsightly buildings, spill over the tops of walls, rapidly form an effective windbreak when planted 3 to 4 feet apart and are perfect companions for small shrubs and perennials. Even when not in bloom, the fine to medium textured leaves are attractive in all seasons, giving the plants a billowy appearance.

Designing with Oleanders

Despite its known toxic qualities, the oleander remains one of the most popular ornamentals for warm climates. Bob Newding, in our interview, gave us extemporaneously some of the most illuminating insights into designing with oleanders, particularly in Zones 9 and 10 but applicable to seasonal planting in cooler zones as well:

First of all, I think lots of people tend to think about oleanders in monoculture, either as mass plantings along highways or along a border of a parking lot, or maybe as accent plants you might find in someone's yard, and that's it. I like to think of oleanders as part of mixed plantings, mixed with other evergreens, both hardy evergreens and some tropical evergreens. And in doing so, choose the oleanders so that you have in

Oleanders in Hawaii with cloud covered mountains providing a dramatic background.

Freeway plantings of oleanders (lower left).

Galveston home with oleanders planted bordering the roadside.

the painting . . . tall background plants, plants that tend to be upright, plants that are intermediate-size in the foreground, maybe spreading or weeping varieties, and in the very near foreground possibly having some of the petites and mixing those with hardy evergreens like Pittosporum. And if you want color contrast with a lighter gray-green, there's Eleagnus; if you want texture contrast, the spreading form of the Coppertone Loquat that comes out of California, another Monrovia plant with large, dark green, waxy leaves that tends to mound on itself and has copper-colored new foliage; Sago Palms; even bulbs planted with various ground covers and annual or perennial flowers; Vinca, the Periwinkle, which is in the same family as the oleander and comes in various colors and can compliment the blooms [of the oleander], mixing colors [of oleanders] that go together and in some cases contrast with each other but are broken up by the textures and colors of the other plants that you put with them.

I also mix in plants that have fragrance; the oleanders that have fragrance I mix also with Night-blooming Jasmine and, of course, the Carnation of India is fragrant, especially in the morning when the blooms are fresh they smell somewhat like gardenia. The idea is to use the oleander in combination. If you have a major freeze or you have a problem with disease you can lose a lot of your oleanders, but if you use plants that have different hardiness, different bloom times, you've always got some beautiful oleander that's blooming as an accent, as part of the collage. And even in the winter when they're not blooming, you have the difference in color, texture and the size of the leaves, for instance, in combination with other plants. If one would take the time to study oleanders, [one would find that] oleanders are relatively inexpensive, relatively easy to grow, relatively easy to prune and maintain. I think one of the preferable ways to grow oleanders is in combination with other plants.

Oleanders are commonly seen as hedges used as screens around a yard, as windbreaks in places exposed to constant and often strong winds and to reduce traffic noise, but they are superb when planted as single specimens or as tree forms. (See section on Pruning in Chapter 8.) In fact, they are so versatile that many landscape formats such as massed plantings of mixed species incorporate oleanders of different sizes and flower colors, and on slopes and hillsides they are especially desirable for their full, bushy habit and their fibrous root system which provides excellent erosion control. Of course, for hedges, windbreaks and sound barriers, choose only the most cold hardy varieties.

Ted Turner, Sr. tells us that much of the landscape design currently in vogue in Corpus Christi consists of starting with larger plants as background, then scaling down to an intermediate size and finally to a ground cover type in front. Thanks to the Turners' work in developing more varieties with smaller growth habits, oleanders can play a greater role in this type of landscape design. When the free-blooming petites are mixed with other plants, especially evergreens, and used in the foreground, they are never overbearing and add a splash of color during the summer when other plants are past peak bloom.

Comparison of Apple Blossom (left) and East End Pink (right).

Oleanders grown as a hedge with each plant
pruned according to its nature.

Recent curbside plantings of oleanders in Galveston
- already in splendid bloom.

An important observation Bob Newding makes is that oleanders, although salt and wind tolerant and exhibiting no problems growing around Galveston Bay or along the estuaries, often incur serious wind and salt burn on their leaves when planted along the ocean front. He illustrated this by taking me to beachfront plantings that were most unsightly because the majority of their leaves were burned and brown with one-third to one-half of the leaf affected. To alleviate this he suggests the following landscape technique:

> In planting oleanders on the ocean front with prevailing breezes and salt mist, they must be protected in some way. And, believe it or not, it doesn't mean they have to be planted behind something. Oleanders planted on the ocean front that are right in front of buildings do fine because the wind doesn't blow through the leaves. The physics of the shape of the building behind it changes the wind flow such that the plant is fairly protected. Another way that I recommend is to plant salt tolerant evergreens in front [on the ocean side] of the oleanders such as Pittosporum, for example, then plant the oleanders just on the other side, or downwind from them. The oleanders will actually go up higher than the *Pittosporum* so you can see the blooms, but those hardy evergreens will change the flow of the wind so that it comes up over the top of the plants behind them, thereby protecting them. Any time you can create a ramp of plants in front of an oleander planting you will be redirecting the wind flow.

Companion Plants

In addition to the "Family Members" mentioned in Chapter 1 and the suggestions by Bob Newding, there are many possibilities for companion plants in warmer climates. *Lantanas* — a good choice would be a variety called 'New Gold' for its exceptional flowering qualities — set among rocks around a multi-trunk oleander create a stunning sight with both plants in bloom from spring until fall. Ground cover *Sedum spp.* are also extremely attractive. In flower beds, street medians and large planters there are many low-growing annual flowers that may be used to create a carpet of any hue to harmonize or contrast with the color of the taller oleander. An idea that is often overlooked is the use of flowering vines as groundcovers. Compact, shrub-form *Bougainvilleas* make an especially good choice as they will not tend to climb the stems of the oleander and have similar cultural requirements. Literally hundreds of perennial plants and herbs would make colorful and fragrant companions with the only exception being those with very different cultural needs.

Selecting an Oleander for Your Area

The major goal of oleander hybridizing during the past ten years has been the development of compact varieties, petite and dwarf forms, and free-blooming cultivars, shifting away from utilitarian screens and windbreaks to more ornamental uses.

There are several factors to consider when choosing an oleander for your garden. If you live in a sub-tropical region, you naturally have a much greater choice because plants can be grown in the ground year round. Nonetheless, ultimate size is still a determining factor, as is growth habit, hardiness, color selection, flower size, length of flowering season and how profusely the plant blooms. If you live in a temperate area, from the cooler areas of Zone 8 to Zone 4, you will probably want a plant suited for container culture. Although many varieties described in the list of selected cultivars in Chapter 5 will grow well in containers, we would recommend beginning with some of the free-blooming petites or dwarfs. Tree form oleanders are successfully grown in containers in colder areas as well, as long as they are given protection from freezes. If you intend to grow a tree form oleander in the ground in Zone 9 or 10, be certain you have chosen a hardy variety so if you do get a freeze your plant will survive. Remember that if a tree form freezes to the ground it will come back as a shrub! One frost-free area where oleanders do not do well is along coastlines where fog is prevalent much of the year. Here growth is generally thin and leggy and flowering is sparse. Oleanders in hot, humid climates with heavy rainfall such as southeast Asia often have a difficult time performing at their best. Interestingly, many sources agree that the double-flowered varieties provide a better display than the single-flowered types in situations such as these where conditions are less than ideal.

In speaking with Dr. Jerry Parsons, I learned some interesting facts about the San Antonio, Texas, area. Since no one can exactly paraphrase Dr. Parsons and do him justice, we quote in full:

> You see, there's not really that much work being done with oleanders. It's a very underused plant, mainly because nobody's interested in it north of here, because it will freeze. Of course, we've got a situation here. A lot of people don't realize it in these urban areas, but one of the big problems we have is deer. The deer population is just terrible. And somebody calls and says, "I want an evergreen shrub for a screen that's green all winter and that blooms in the summer and is drought tolerant. What can I plant that the deer won't eat?" Well, what is that? It's the oleander. And people say, "I don't like the oleander." Well, you better learn to like the oleander, its the only damned thing you've got to plant out there. They'll [the deer] eat the hell out of everything else! . . . they'll eat the first new leaves of an oleander in the spring but they don't eat 'em very long!

Growing Oleanders in the Ground in Temperate Climates

We are fortunate to have the experiences of Clarence Pleasants as guidelines for growing oleanders in the ground as far north as Norfolk, Virginia. During his years at the Norfolk Botanical Garden, he developed a thorough knowledge of oleander cultivation in an area that, during the 1950's and 60's recorded average low temperatures in the upper 20's and rarely dipped below 19°, and then only for a few hours. Freezing temperatures in this range would last for several days and would occur every 2 to 3 years. Oleanders might grow to a height of 6 or 8 feet and then a freeze would kill them to the ground, thereby limiting their ultimate size. As Clarence said, "Mother Nature did the pruning."

In temperate climates full sun exposure is even more critical for plants to flower well. Clarence observed that the best locations were against a wall or building where they were exposed to maximum sunlight and heat and protected from the wind. During the years he collected cuttings from the eastern part of the United States, he noticed significant differences in cultivars with some varieties much more wind tolerant than others and a few displaying greater cold tolerance than those commonly grown in the area. A fascinating report (appearing in a recent issue of *Nerium News*) is from an enthusiast in Villas, New Jersey, who has been growing oleanders for about 35 years and has amassed a collection of several thousand plants. Even more astounding is that he is growing them out of doors in the ground year-round! Located near the southern tip of Cape May County near the Atlantic Ocean and only blocks from Delaware Bay, he finds the location to be a microclimate with conditions similar to Zone 8 or 8B and is also able to grow other relatively tender plants such as palms and gardenias without winter protection.

Oleanders are often grown in large tubs in the Norfolk area, usually planted in heavy clay soil. Gardeners would tell Clarence how much work it was to move plants indoors in fall and out again in the spring. (This is another reason to grow oleanders in lightweight, highly organic soil mixes!) For oleanders grown in the ground, Clarence recommended the addition of organic matter for root development and good drainage but cautioned that in colder areas overly rich soil conditions should be avoided. It was his contention that if plants are discouraged from growing rapidly and putting on extensive growth they will go dormant earlier and increase in hardiness.

Oleanders Grown as Annuals in Colder Areas

Oleanders are excellent grown as annuals in areas too cold for successful overwintering or if one hasn't a greenhouse or sunroom in which to keep them throughout the winter months (or a desire to haul them in and out of a cellar). The dwarf and petite forms are especially good choices as they are very early to bloom and with regular care and fertilizing will continue until the first frost. There are not many exotic plants that will flower for such a length of time and the cost of an oleander is in the same range as many tropical hibiscus which are often treated as annuals in parks and gardens in the north.

*I never saw a garden from which
I did not learn something.
Russell Page,
The Education of a Gardener*

The petite oleander, Carnival, flanked by companion plants - From left *Cestrum nocturnum* (Night-blooming Jasmine) in heavy bud, *Cycas revoluta* (Sago Palm), *Tabernaemontana divaricata* (Crepe Jasmine, Carnation of India), photographed at Bob Newding's home.

Some Sources for Oleanders

Wholesale

Aldridge Nursery, Inc.
Von Ormy, TX 78073

Monrovia Nursery Company
18331 E. Foothill Boulevard
Azusa, CA 91702-2638

Cottage Hill Nursery, Inc.
9960 Padgett Switch Road
Irvington, AL 36544

Reasoner's Tropical Nurseries, Inc.
4610 14th St.
West Bradenton, FL

Hines Wholesale Nurseries
PO Box 42284
Houston, TX 77042

Mail Order

Logee's Greenhouses
141 North Street
Danielson, CT 06239

The Plumeria People
910 Leander Drive
Leander, TX 78641

Retail

Numerous retail nurseries and garden centers throughout the U.S. offer oleander plants during spring and summer.

Franklin D. Roosevelt

"Little flower – but if I could understand
What you are, root and all, and all in all,
I should know what God and man is."
                                    Alfred Lord Tennyson

Oleander Research

Plant research is often highly technical requiring sophisticated laboratory facilities, years of background and training in botanical sciences, and many more years of painstaking experimentation and documentation. Current research is focused on the various cardenolide glycosides and other constituents in oleanders and their pharmaceutical applications, especially with regard to cancer. Known historically to have been used to treat cancerous ulcers and external carcinomas, extensive research was begun on the oleander in the late 1950's by Turkovic and Jäger et al., continued in the 1960's by Janiak et al., in the 1970's and 1980's by Yamauchi, and by Tittel and Wagner in 1981.

The field of taxonomy is also extremely complex involving many aspects of botanical study from plant physiology and phytochemistry to herbarium specimens and field trials. For more details on the subject, we recommend Leeuwenberg's *Series of revisions of the Apocynaceae XII, 1984 and notes on Nerium L.*, F.J.J. Pagen's completion of the revision under Leeuwenberg's supervision, and Pagen's book *Oleanders, Nerium L. and the oleander cultivars (1987)*. Both are published at Wageningen Agricultural University, The Netherlands, and thoroughly cover the taxonomic aspects of the genus.

Our primary focus is on research relating to ornamental applications, i.e. greater cold hardiness, hybridization leading to new colors, more dwarf and compact plants, different fragrances, etc. We hope all gardeners will become "researchers" by collecting and photographing the best varieties, and will encourage nurserymen to propagate them so a greater selection can be made available to the gardening public. One of the most important scientific projects that could be undertaken and should receive grant consideration is a "key" to the named cultivars throughout the world. This would clarify the vast confusion that exists with plants coming from one area of the world and being renamed in another. Although it would be a costly and time-consuming project, possible only with the aid of gene mapping, similar research is currently being done on other plant species to establish irrefutable documentation. Bob Newding had this to say on the subject of nomenclature:

Well, I'm a trained biologist so there's a certain psyche that biologists have about things. If you collect one you want the next one, and the next one, and pretty soon you want to complete your collection. After visiting with people who knew the nomenclature I started going around trying to name the ones that I saw and oftentimes I'd make mistakes as I would see something that didn't look the same as it did a month earlier. I found out that oleander blooms change their characteristics as the year goes by due to exposure to heat, wind, etc. Some of the flowers and some varieties simply change color from the time they first bloom until they finish blooming at the end of the summer. The 'Franklin Delano Roosevelt', for instance, starts out as a salmon pink and it goes almost to an orange, and at the end of the summer it's copper colored. So, unless you see it at different times, you'll think it's a different plant.

The Turners in Corpus Christi are presently concentrating their efforts on creating a true lavender or purple oleander as well as developing a new category which they have tentatively entitled "miniature" or "toy." Plants in this group would not exceed 2 feet in height and would, of course, exhibit all the characteristics of the best oleanders. All creative challenges contain inherent risk, however, and the Turners developed their first miniature in 1983 only to lose it in the aftermath of a storm. It was an ideal hanging basket plant with double pink flowers on cascading branches. We mention this to inspire anyone interested in hybridizing to begin — there is much more to be accomplished in the world of oleanders. As can be seen by the dates of introduction of new varieties from the Turners and Monrovia Nursery Company (listed in Chapter 5), some of the finest cultivars have been introduced in the past several years and more are still to come! (See also Chapter 14 on Plant Patents.)

If current research through hybridization or by the exchange of plants is successful in developing more cold hardy varieties, then oleanders can be grown in areas other than Zones 8, 9, and 10 and their popularity will increase dramatically. Three or four years ago Dr. Jerry Parsons and Greg Grant (of Lone Star Growers in San Antonio) visited members of the International Oleander Society in Galveston and were given cuttings of the five hardiest selections that had survived numerous freezes. Dr. Parsons notes that until this time they only knew of the cold hardiness of 'Hardy Red', 'Pink Beauty' (aka 'Hardy Pink') and possibly 'Little Red', varieties that survived the 1983 freeze in San Antonio when temperatures plummeted to six degrees and remained below ten degrees for many days. As Dr. Parsons commented, "It was terrible. It wiped out the entire citrus nursery business in Texas — right to the ground. It took the nursery industry to its knees." The cuttings given to Dr. Parsons and Greg Grant were rooted and have been grown in the field for several years. After several winters they plan to locate the survivors farther and farther north to continue the selection process.

Additional research has been undertaken by Dr. Jerry Parsons and Dr. Wayne McKay in El Paso. Using seeds supplied by Kewpie Gaido and Elizabeth Head that Dr. Parsons had irradiated at Texas A&M University, their goal is to determine whether there are any oleanders superior to industry standards in terms of color, dwarfing, cold tolerance, etc. Approximately 600 seedlings were planted at the experimental station in El Paso and excess seedlings were taken by Dr. Parsons and planted at Lone Star Growers in San Antonio. In El Paso normal winter temperatures are at the edge of the cold tolerance range for most oleanders (Zone 7B), but Dr. McKay pointed out that until this year (1995-96) there has only been one severe winter when the temperature dropped to 10 degrees for one night and into the low teens for several nights. All plants were examined to determine freeze damage. No plants were killed to the ground but any plants with burn-back of the tops were destroyed. This eliminated approximately half of the total number. The remaining plants did not show any signs of winter injury. The seedlings are approximately four years old and most flowered in the first year. Plants range in size from dwarf to large and have exhibited almost every shade from white to red, including salmon and pink, and several that appear new and of exceptional color. In the near future Dr. Parsons and Dr. McKay will propagate any promising plant to test it on its own, and secondarily, collect seed from that plant to determine the results of the irradiation on its progeny.

This note on irradiating seeds is offered by Dr. Parsons:

> We found out that when you irradiate a seed, you can possibly get mutations or "sports" from the initial planting of seedlings, but the best chances [for mutations] are when you collect seed from those [irradiated] plants. Out of all those hundreds of seedlings, he [Dr. McKay at El Paso] took the ones he could plant in his block — when you're dealing with an experiment station, a research station, they have to have a certain number of plants when they plant their plots; in other words, there have to be exact numbers of every kind of treatment. I took the rest of them, which were just throwaways — I have this thing about throwing away plants — and put them out here at Lone Star [Growers], and it's amazing the differences we have. You can go out there and walk up and down the line and every single one of them is different.

We have reports that some exceptionally cold hardy plants have been developed in Israel and Italy. It would be an exciting project for a group or individual to perform a series of controlled crosses between these varieties and some of the most promising cold tolerant American varieties.

When at last I took the time to look into the heart of a flower,
it opened up a whole new world . . .
as if a window had been opened to let in the sun.
Princess Grace of Monaco

117

Magnolia Willis Sealy

No occupation is so delightful to me
as the culture of the earth, and no culture
comparable to that of the garden.
Thomas Jefferson

Galveston, the Oleander City

More than any city in the United States, Galveston, Texas, has treasured the oleander. In 1841, Joseph Osterman, a prominent businessman of the day and a merchant and ship owner, brought the first plants from Jamaica as gifts for his wife and his sister-in-law Mrs. Isadore Dyer. These first oleanders, a single white, and a double pink later named after Mrs. Dyer, were planted on the grounds of their homes. Mrs. Dyer loved the flowers and propagated many plants, giving them to friends throughout the island.

Plants and cuttings were also brought from the Middle East and Europe by prominent Galveston families. There is a reference to a Dr. Roehmer, visiting from Germany, who was impressed with the roses and oleanders in bloom in 1846. In 1900, a year sadly remembered by Galvestonians, much of the island was destroyed in The Storm, a severe hurricane that brought gulf waters as high as the first story of many of the original homes and public buildings. After the storm, as a precaution for the future, tons of soil and sand were brought in to raise the level of the island covering virtually all remaining plant life. With Magnolia Willis Sealy as its inspirational head, the Women's Health Protective Association was formed to improve health conditions and beautify the city. In the early days in Galveston, especially in the aftermath of the storm, fragrance was of major importance. The Women's group chose the oleander for planting in parks and public gardens and to line the streets and roadways, believing it to have healthy properties in addition to the strong scent which would drown out all the noxious odors from ponds, stagnant water and sewage problems. It is reported that the fragrance of the original oleanders was so intense that people would often have to close their windows at night! Oleanders were also recommended for planting around the perimeter of yards and vegetable gardens to serve as windbreaks.

In 1908 the *Galveston Tribune* commented that together with the beaches the city's oleander plantings were ranked as one of the island's most popular tourist attractions. Since 1910, when the *Galveston Daily News* reported that Galveston was known throughout the world as "The Oleander City," parades, balls and other festivities have become a tradition honoring this lovely and enduring plant that has contributed so much to the beauty of the island.

Dr. Darrell MacDonald conducted extensive studies on the flora of the island for his PhD dissertation which focused on the transformation of Galveston from a barrier island into an urban complex of exotic plantings mixed with native plants. His research involved 120 gardens located in four sections of the island with 30 in each section so that he would have a statistical set. He censused and mapped each garden in addition to interviewing the owners. Dr. MacDonald came to the conclusion that the main identifying factor is what he calls the Triad — the oaks, the oleanders and the palms. "That group of plantings, I think, is really an historical trademark of the move from a barrier island to a tropical-like environment which they wanted their city to be," he stated.

Many of the original oleanders introduced into Galveston have undergone several name changes as new names were often given to honor prominent Galveston citizens. Oleander historian Elizabeth Head believes that some of the older varieties match pictures she has seen of European ones. On an oleander list of the 1920's both 'Early Dawn' and 'Peace' are listed, names that she feels are much prettier than people's names. In a 1941 compilation sixty varieties were listed, but by this time plants that had appeared on previous lists had undergone many name changes.

Children grew up among the plants and flowers on the island and whenever we would ask how someone became interested in oleanders there would be a noticeable glint in the eye as the years rolled back and fond memories of childhood returned. Some of the most memorable moments in our interviews would be the recounting of these childhood experiences by oleander aficionados in Galveston. In her interview Elizabeth Head recalled, "I became interested when I was very young because I'm a 'born-on-the-island' person with oleanders all around, . . . and I've always loved them and enjoyed them in my neighborhood. We used to string them and wear them around our necks and for the number of children and oleanders in Galveston, there aren't any reported poisonings." Bob Newding remembers, "I grew up on the island and oleanders have always been a part of our life growing up. They were around the elementary school that I attended, around the junior high school, and we used to make necklaces of the blossoms for our girlfriends when we were kids, 10 or 11 years old — we threaded the flowers and made leis. Galveston, growing up in Galveston, having it called The Oleander City, they've always been planted around, especially the municipal buildings and neighborhoods and along the alleys. Galveston's unique in that it has alleys between the main streets and these are areas that get sunshine but yet they're protected; . . . as a kid growing up, this was a safe place to play, a kind of grotto or retreat, and some of the most beautiful oleanders in Galveston are planted alongside the alleys. I can remember that as a boy, somewhere from the mid 50's to the early 60's, there have always been various individuals or government agencies that have tried to promote oleanders in Galveston."

Clarence Pleasants devoted his life to all aspects of the oleander, especially research into their history and culture. His observations were detailed and extensive as was his knowledge of cultivars. Professor Octavia Hall says that he could "sense" a particular variety unlike anyone else. Of her work in putting together oleander herbarium specimens Professor Hall wrote (regarding the morphological variations constituting varieties): "In these Clarence's exquisite ability to spot subtle variations was peerless. I deferred to his judgment all the way, giving him credit for variety definitions in each of the herbarium sheets." One can only hope that some-time in the not too distant future, the International Oleander Society, or perhaps the Galveston Chamber of Commerce, will secure the necessary funds to reprint the classic book first published in 1966 by Clarence Pleasants entitled, *Galveston, The Oleander City*. When we interviewed Clarence on tape he gave his permission to quote anything we wished from his writings and experiences, such was his generos-ity and selflessness. We would like to include a brief selection from his book as an example of his love for Galveston:

". . . as a gardener I could not fail to be attracted to Galveston through its jewels, the oleanders; but aside from that, I admire and praise this city that has trod the sands of time alone and stands today as a monument to a gallant people. . . . Truly one would have to visit this city of floral beauty to understand why oleanders and Galveston are synonymous."

Where to See Oleanders in Galveston

Moody Gardens

Gazing across Galveston Bay from the Texas coast, visitors are often surprised by the unusual sight of a huge glass pyramid rising from a beach of pure white sand. The Pyramid is, in fact, a ten-story, 40,000 square foot greenhouse enclosing a fully functioning, chemical-free rainforest kept in balance by beneficial bacteria and in-sects, and several dedicated horticulturists. Paths guide visitors past rocks and wa-terfalls amid a superb collection of tropical plants, several species of tropical birds and reptiles, and more than 2000 colorful butterflies that glide weightlessly among branches laden with orchids, ferns, plumeria blossoms and flowering tropical vines. It is one of the central features at Moody Gardens, located at One Hope Boulevard on the north shore of the island. The project also offers a conference and convention center, horticultural therapy and hippotherapy programs, entomological and medi-cal research centers, and an environmental library as well as a hotel, restaurants, an IMAX-3D theater, volleyball courts, paddleboats and other activities. Today, with over 20,000 plants — hundreds of which are oleanders — on the 156 acres surround-ing The Pyramid and other buildings, the gardens are an oasis of lush green and rainbows of color, and are becoming known as one of the best botanical collections in the southern United States.

John Kriegel, Director of Gardens, was of great assistance in providing a com-prehensive overview of the focus of the gardens as related to oleanders. With the aim of establishing a collection worthy of Galveston's reputation as The Oleander City and the home of the International Oleander Society, they began with a group of about 60 varieties amassed and donated by Clarence Pleasants. With these they established an "I.D." garden, or world reference garden, for the purpose of identify-ing cultivars as well as public enjoyment. It is in this first of many planned "I.D."

Massed oleander planting with the Moody Gardens Rainforest Pyramid in the background.

Oleanders, pittosporum, periwinkle and palms in an attractive and harmonious planting at Moody Gardens.

Massed oleander planting at Moody Gardens.

gardens, now known as the Clarence Pleasants Oleander Collection, that there rests a plaque dedicated to his memory. It reads:

> "Inspirational Founder of the
> International Oleander Society,
> Author, Teacher, Horticulturist,
> and a Southern Gentleman"

Larger quantities of ten or fifteen selected varieties have since been propagated and integrated into mass plantings in the landscape. As Kriegel says, "Basically, utilizing oleanders wherever we can has been our design criteria." Future plans include accession of varieties from other parts of the world as a continuation of Clarence Pleasants' collection and identification work, propagation for landscaping around a new hotel and smaller pyramid that will be a Discovery Museum, propagation of unusual varieties for distribution to the public, and making the oleander collection more informative and educational as well as beautiful.

One of the finest sets of herbarium specimens we have seen is the oleander collection prepared by Professor Octavia Hall. Her work is meticulous and artistic and includes stems, leaves, flowers, follicles and seeds. The collection is kept in a fireproof safe at Moody Gardens. A duplicate set was submitted to and accepted by the Smithsonian Institution.

Open Gates — The George and Magnolia Willis Sealy Conference Center

The Open Gates garden is located on Broadway between 24th and 25th Streets and was dedicated by Eugenia Sealy Cross, George Sealy III and Lane T. Sealy in cooperation with the University of Texas Medical Branch, the Magnolia Willis Sealy Foundation and the International Oleander Society in memory of George Sealy, Jr., 1880-1944. George Sealy, Jr. and his mother Magnolia Willis Sealy are remembered for their lifelong dedication to oleanders in Galveston. Open Gates was the original family home and more than thirty varieties of oleander as well as many other flowering plants may be viewed in the newly landscaped gardens.

Notes and Anecdotes from George and Lane Sealy

The first generation of the family in Galveston were George and Magnolia Willis Sealy. Magnolia was from the little town of Montgomery north of Houston and George had come from Pennsylvania. As George Sealy III related to us:

> In 1900 or 1901, after the Galveston storm, grandfather was on his way east to refinance Galveston's public debt and he died on the train going north. Since Magnolia was a widow for the next [approximately] 30 years, she traveled a lot around the world and brought seeds and plants back to Galveston from everywhere. We had lilies from the Nile Valley in the yard and palm trees from all over the place and flowers from any place she could bring seeds. She loved flowers. Grandfather's

name is with hers on a stone down in the Bolivar Ferry Park which mentions that they were instrumental in putting palm trees all along Broadway so I presume the idea started before granddaddy died. But, nevertheless, she continued it. She was quite active in flowers. On the northwest corner of the old family home on Broadway there's a plaque to the 'Isadore Dyer' [variety], right across the alley from our home. Because she was attributed with starting the first real interest in oleanders on the island, I presume that was where they thought the first oleander was planted. I imagine because it was so close to the house, probably grandmother took great delight in it since it blooms so much and so well and so beautifully in Galveston. She was quite helpful in spreading them, too. But from what I remember and from what I remember daddy telling me, and mother, too, dad [George Sealy, Jr.] was the propagator — that was one of his real serious hobbies. He was trying to find something that would grow in Galveston with the salt breezes and he tried everything. He never quit experimenting with flowers to see if something would do better than oleanders in Galveston but nothing did.

George Sealy, Jr., like his father and mother, was in many ways very influential in the development of Galveston. He was president of the Galveston Wharf Company that operated big shipping docks and grain elevators. The company also serviced the deep water shipping business at the turn of the century when Galveston was the only deep water port except for New Orleans and New York. Cotton was the big crop then and the company warehoused cotton, sulfur and other materials until they were loaded onto ships. Another of Sealy's business interests was a Cotton Concentration Company which transported raw baled cotton into Galveston by rail and, using huge hydraulic compresses, compacted it into small bales that required less space in the holds of ships. Of these years with his father, Lane recalls:

As his hobby he and Eddy Barr raised oleanders from seeds [from plants he pollinated] and he did grafting also. His "little" oleander operation was on the grounds of a vacant lot of the Cotton Concentration Company yards. It was sort of in a flood plain, marsh land, and subject to backup from the bay. There would be acres after acres of little buckets with little twigs in them and Coca Cola bottles all over the place with twigs in them. His idea was to be able to raise every known variety of oleander on the island and make those available free to the public. He wanted to plant them without charge for the city and did — the city is just profuse with them. He and his mother were both responsible, much to the chagrin of the city that has to maintain them!

I remember as a boy of perhaps seven or eight during the war when we would have blackouts and coastal defense drills, he would go out at night and find a variety of oleander somewhere in town in somebody's yard and he would steal them. Well, I hate to use the word steal. What I mean by stealing is he always had a little shear in the back of his car, pruning shears, and he would go up to an oleander plant that he wanted

124

George Sealy

and take a clipping from it. He was so fearful of being caught doing this that he would entice me to. . . . I guess you really had to know my dad. He was the epitome of honesty and integrity. The very thought of anything off color was just abhorrent to him and he raised us that way. So he'd stay in the car and show me where it was and stay a couple of blocks away and make me go take the oleander clipping. I thought this was great sport — a big old tree there and we just took a little clipping off of it. It turned out later that when I told this story to the Oleander Society everybody whooped because it seems everybody does that. But to me it was in the dark of the night and we were really doing something evil. I was perpetuating this sinister operation of dad's and I felt somewhat guilty about it. I remember when he developed new varieties he named one after my mother and others after very prominent people. Once he said he had a new variety and couldn't think of a name for it and — I remember this so distinctly — I said, "Well, look Daddy, it's the least you can do since I have to do all your undercover work and strategic stealing of these oleander cuttings half of my life, the least you could do is to name one after me." All in jest, I remember that distinctly. "I've been doing your dirty work for you and you just owe me that much of a favor to name one after me." Of course, being a youngster of seven that's the last I heard of it and never thought any more about it.

Then, within the last five years, long after dad and mother died, my brother and sister and I wanted to give some kind of memorial to Galveston that could be used and enjoyed by the whole city in their

Oleanders at Open Gates, Galveston

(l to r) - Algiers, Ed Barr, Pink Beauty

Oleander plantings at Open Gates in Galveston

Oleanders at Open Gates, Galveston

Oleander plantings at Open Gates - right to left - Mrs. Robertson, Petite Pink,
Mrs. F. Roeding, Sue Hawley Oakes, General Pershing, Mrs. Robertson (above),
Petite Pink (below), Pleasants Postoffice Pink, Turner's Shari D.

Oleander plantings at Open Gates in Galveston

memory, so we thought of doing something with oleanders. Well, we didn't know anything about oleanders, so we just thought it would be a good thing to make a park or a memorial of some kind. Then my brother and sister said, "Let's see if we can't donate a piece of ground and do what dad did, just develop oleanders and give them away, without fanfare or money or anything." Dad would give them to anybody. The only agreement was that they would plant it, take care of it, nurture it along and try to develop more of them. He sent them all over the world. So we wanted to do something like that, create something so that people from all over the world could get an oleander from Oleander City and take it home. Well, we didn't have the foggiest idea where we'd start but my brother George said, "Hey, I saw this guy on television who was called Mr. Oleander and he seemed to know everything there is to know about oleanders. Let's go to Galveston and find out who he is." So we called the station and his name turned out to be Clarence Pleasants. We called him up and he thought we were a couple of screwballs asking a bunch of questions. Of course the name was very familiar to him but to have two people sort of come out of nowhere into his life of oleanders this late in the game, something just didn't smell right to him. Of course, he didn't put it quite like that — he just didn't know if we were legitimate or not. But he answered very genuinely, "Yes, I'd be glad to talk to you about oleanders and a memorial for Mr. Sealy and whatever else you want to do. You came at a most propitious time, it just so happens that our local Oleander Society is having a meeting at Kempner Park today and we would love for you to come." (That very day we were there!) We said, sure, we'd be glad to. We showed up and introduced ourselves to the ladies and were treated royally because we didn't know they existed and they didn't know we existed. We were just sort of something out of the past that popped up and said, hey, we want to try to work with you all on this memorial that we have in mind. And they, of course, perked up right quick and said, "How many ideas do you want?" That's how our relationship with the Oleander Society really got started — a phone call out of the blue.

But how this ties back in to my sleuthing career was that Clarence said, "You're Lane Taylor Sealy, aren't you?" I said, "Yes, but how in the world did you know my whole name?" (I don't use the middle name Taylor.) And he pulled out his list of varieties of oleanders and said, "Look here, there's a 'Lane Taylor Sealy' oleander." I thought, well, I'll be damned, my dad did it. I never knew, from all that time long ago when I said this to dad that he really did do it, and it was an official variety in the books of the oleander world.

Later, when the Sealy family was working with the landscape architect on the new design for the grounds of Open Gates, Lane contacted Julia Allison, then president of the Oleander Society, about acquiring oleanders for the project. In his conversation with her he made this request:

"... there is one thing that I would like especially to do if you could get permission from somebody — my days of stealing are over now — but if you could get permission from whoever now owns the property of SeaArama to take a clipping off of that 'Lane Taylor Sealy' variety and start one, I would be most grateful." That was the only known 'Lane Taylor Sealy' variety in existence as far as any of them knew. So Clarence outdid himself. He went and dug up the whole bush and put part of it in Kewpie's greenhouse, and then planted one in Open Gates and it was appropriately labeled. A year after that when it was all done we went down and dedicated it, had a beautiful little ceremony in the garden there. I gave the address on behalf of the family and when I finished they said they would like to give me a gift. They gave me the most beautiful plant of 'Lane Taylor Sealy' that Clarence had raised in Kewpie's greenhouse from a clipping he had taken maybe a day or so after we first met. It had grown to maybe eighteen inches high in a pot and it was so

Open Gates, Galveston

beautiful. I was thrilled to death, holding back the tears, I was so touched with it. Well, I said I can't take this to San Antonio, every time I try to plant one I kill it with kindness. I just don't have a green thumb — we stuck these things in Coke bottles like I thought my dad did and his grew and mine didn't. So I asked them to keep it there in Kewpie's greenhouse and they said they would. As far as I know it's still going in the hothouse and they're making more little Lane Taylor Sealy's from it!

Beauty is God's handwriting.
Welcome it in every fair face,
every fair sky, every fair flower.
Charles Kingsley

129

Maureen Elizabeth 'Kewpie' Gaido

Clarence Grant Pleasants

November 11, 1916 - August 19, 1995

April 4, 1930 - December 27, 1995

12

*He who is born with a silver spoon in his mouth
is generally considered a fortunate person, but his
good fortune is small compared to that of the
happy mortal who enters this world
with a passion for flowers in his soul.*

Celia Thaxter

Distinguished People

Who Have Contributed to the
Worldwide Appreciation of Oleanders

We begin this section by honoring two people who have been the forerunners in advancing the cause of oleanders; "Kewpie" Gaido, to whom this book is dedicated, and Clarence Pleasants, "Mr. Oleander" to those who had the good fortune to know him and work with him.

Maureen Elizabeth "Kewpie" Gaido

A special friend who, through her energy and enthusiasm, inspired everyone she met to share in her great love of oleanders. Kewpie was born in Galveston, Texas, and devoted much of her life to promoting oleanders there and elsewhere throughout the world. After being inspired by Clarence Pleasants, she was instrumental in founding the National Oleander Society (later changed to the International Oleander Society) and was its first president from 1967 to 1972. Her knowledge of oleanders and her ability to speak eloquently of them has encouraged people all over the world to grow them. Her love of flowers led her to travel widely, visiting gardens everywhere seeking to increase her knowledge of plants. She visited Greece, China, Japan, Copenhagen, the French and Italian Riviera, and Monaco where she saw extensive plantings of oleanders. While in Italy, on a visit to Luca, she saw immense oleander trees balled and burlapped at a nursery and noted that temperatures in this area often fell much below freezing. In Holland, F.J.J. Pagen took Kewpie, Clarence Pleasants and Elizabeth Head to a nursery where he showed them the Italian Oleander of which Kewpie remarked, ". . . they were very beautiful, very large (flowers)." She corresponded with Ronald Reagan in 1971, when he was governor of California, after reading that he had designated many miles of freeways for planting oleanders and sent him plants and cuttings of some of Galveston's finest cultivars. She also spoke with Lady Bird Johnson after hearing that the former First Lady was planting oleanders in other areas of Texas. She worked tirelessly to promote oleanders in Galveston as well, saying of the plant, "You're never a hero in your own town," and encouraged local residents to understand the treasure that Galveston offered to the world. Dr. Darrell MacDonald said of her, "Mrs. Gaido's altruistic dedication to the oleander and the idea of beautifying the island got me interested in trying to help them solve the problem of all the varieties." He added, 'She was such an important force in making people realize that the aesthetics of the island are as important as its economic functions." She took great pride in the "international" aspect of the Oleander Society, commenting that there were members from eight countries. In 1975 she was made an honorary life member of the society and was presented with a plaque of appreciation as organizer and first president.

Kewpie was well known to Galvestonians as the matriarch of the island's famous restaurant family, and as a native Galvestonian herself was intimately familiar with the grand social events of the past. She told us of the Oleander Festivals of former years when a Queen of the Oleanders was selected along with a train of "little princesses" and floats were decorated with thousands of oleander flowers. Kewpie was instrumental in reviving the Oleander Festival through the International Oleander Society in 1989. As Chairman of the Festival in 1991, on learning that Van Gogh had painted an oleander, she arranged for Dr. Sweetman, author of *The Love of Many Things, A Life of Vincent Van Gogh,* to visit from England and Dr. Frank Pagen, author of *Oleanders, Nerium L. and the oleander cultivars,* to travel from Holland to lecture at the Oleander Festival. She never ceased to be amazed that from a barren island the early settlers created a place of such beauty. Most of all, Kewpie was grateful for Moody Gardens, a vision of such epic proportions that, at first, she thought it would never be realized. She was the earliest champion of Moody Gardens, first president of the Friends of Moody Gardens and Chairman of the Board. She worked unceasingly to instill her love of horticulture in family and friends, taking plants to children in schools to awaken their interest as hers had been awakened by her father. Looking back on her journey with oleanders and all the memories of a lifetime lived to the fullest, she never forgot to mention Clarence Pleasants who she said, ". . . has made all this happen . . . he's a gentleman, very sweet and caring, interesting and interested. This man is a selfless man, very hardworking. He's just an inspiration to us all."

As in her life, so in her death Kewpie wanted her friends to be happy and enjoy themselves. At her funeral, in accordance with her instructions, the Olympia Jazz Band played dirges as well as lively music so that everyone would have a good time celebrating her passing as she wished.

Clarence Pleasants

"Mr. Oleander", as he was known to all in Galveston, saw his first oleanders in Rockbridge County, Virginia, when he was just a boy working as a garden helper. The bright red flowers, growing in tubs on either side of a porch, caught his eye from a distance (the way he says most people often see oleanders). He went to the door and asked the owner of the house what these unusual plants were and she replied, "Oly Andys." From that moment he knew that this was the plant he would study for the rest of his life and dedicated himself to learning everything he could about its culture and origins.

Pleasants Postoffice Pink

The youngest of four brothers, Clarence attended a segregated two-room black school up to grade 7, there being no further grades taught. From her talks with Clarence, Prof. Octavia Hall wrote in 1966: "It [the school] had a broken down shed off the basement, where piles of old *National Geographics* lay helter-skelter, from which he was given armfuls and which he read avidly about a world he knew he'd likely see little of in his life."

Clarence studied under Frederick Heutte, a well respected plantsman, at the Norfolk Botanical Garden in Virginia where he worked for thirteen years. He later developed correspondence with people all over the world in search of information to further his knowledge of how oleanders were cultivated in other countries, the varieties that were available, the forms and fragrances of flowers, the sizes and hardiness of shrubs, etc. He also began a vigorous personal campaign to promote oleanders and traveled throughout the Southeast collecting cuttings that were unusual, once finding the cultivar "Commandant Barthélemy", a double red, growing on the outer banks of North Carolina. During his years in Norfolk he introduced plants from California and many other areas, growing the more tender varieties in the greenhouses and planting unique cultivars throughout the city. Whenever he found an unusual cultivar, whether growing in a yard, on a side street or in front of a public building, he would get permission to collect cuttings and then, by first talking with people about the history and origin of a specific plant, would work tirelessly to identify the variety.

This often led him into some interesting situations. While still in Virginia he corresponded with Mr. Curtis of Seaside Nursery in Galveston. "Mr. Curtis," he recalled, "was most interested in telling me about the oleander and he even agreed to send me some." Further help came from Dr. Snodgrass, a member of the Camellia Society. Clarence wrote to him also and told him "about the oleander and my interest in it, and he wrote me a long letter — busy people usually do — about oleanders. He mentioned that he lived over near the ship channel and had several oleanders, and that 'Sealy Pink' was his favorite." Dr. Snodgrass' letter inspired Clarence even more and he made a trip to Houston, his first flight, to see the oleanders in Galveston. He told a story about this experience that we're sure will strike a chord with plant collectors everywhere:

> And in my collecting [in Galveston] I met some interesting people. I met a gentleman near Kempner Park and I was looking at his oleanders and I told him of my interest in oleanders and where I was from. "Well," he said, "oleanders won't grow up there in Norfolk." I said, "Yes, they do, they do quite well. It's as far north as they grow without needing to be moved in and out. And I'd like to take some unusual plants back." "Well," he said, "you can do that. I can't open the gate because I have it locked (an iron gate), but you can get over the fence." He gave me 'Mrs. Wilson' and this one and another one, about 20 cuttings. He was a very formal man, frightening to a degree, kind of scary until you got to know him. At the end of my visit he was very friendly and nice. He said that he had relatives in Detroit and sent them oleanders for container plants. He even helped me back over the fence! Over the years I'd go out there occasionally and get some cuttings. He once pointed out a plant and said, "This is named after my wife. Her name is Lucille Hutchings." The gentleman was Mr. Sealy Hutchings. He's related to the Sealys and he's kin to the Moodys too. It was a little history lesson walkin' through there.

Clarence concluded with a scene familiar to many of us: "But I did have an enormous amount of cuttings and people were watching as I was walking down the street to the motel. They were sort of surprised because, I mean, what's so exciting about oleanders, why is he doing that? Well, in plumeria country, I'm sure when you're seeking a certain plumeria, the folks settin' on the patio or the upper balcony are wondering what's the great interest?"

After returning to Norfolk, Clarence gathered together all the information he had collected over the years and decided to write a book. He said, "A book would be nice to bring interest to the oleander. The book is not of a scientific nature, it really wasn't intended that way, but it brought interest to the oleander." In addition to his book, Clarence made numerous radio and TV appearances to promote the oleander (much as Kewpie Gaido was doing in Galveston), dispelling the ingrained prejudices people had about their poisonous qualities and pointing instead to the numerous plants grown in parks, along esplanades, and in gardens throughout the Norfolk area. He spoke eloquently on the oleander's many attributes in the garden; its showiness, hardiness, carefree habit and the fact that it is a truly unusual plant that stands out in the landscape. Although it wasn't easy, he recalled, due to the negativity about the oleander's poisonous qualities, he became its spokesman, gaining the respect and admiration of his peers. In his efforts to further promote the cultivation of oleanders, Clarence sent cuttings to the West Indies, traveled to the

Dominican Republic where he presented the Botanical Gardens with 20 of the best varieties in his collection, and corresponded worldwide with horticulturists interested in the genus.

On July 1, 1969, Clarence moved to Galveston to take a position at SeaArama where he began an extensive oleander collection, even designing an Oleander Trail. Prof. Octavia Hall recalls a walk she took with Clarence around the (then closed) SeaArama gardens on July 2, 1991, an anniversary of when he had first started to work there, saying: "He visited personally with each surviving shrub (as I closely observed) that he'd planted there, remarking which had survived and which had not over the years. His intellectual and emotional involvement with his plants was total, as was his involvement generally; and which evoked a like involvement in his friends toward him."

Thus began the wonderful collaboration with the people of Galveston that would last until his death in 1995. Elizabeth Head called Clarence the "soul" of the Oleander Society and his contributions to the city on behalf of oleanders were immeasurable.

Sherry Brahm

Former president of the International Oleander Society whom I had the pleasure to meet on my first visit to Galveston when she invited me to address the Society in 1986. A wonderful person and dedicated leader whose message as incoming President we are honored to quote:

"Those of you in other places can help by letting us know of your experience with oleanders and sharing information of success as well as failure. Help us be a gathering point for knowledge regarding oleander culture."

Dr. Barry Comeaux

Past president of the International Oleander Society and former Director of Horticulture at Galveston College. Through his expertise and technical training in the field of horticulture, he helped the Nomenclature Committee develop a Descriptor List for the identification of oleander cultivars. Dr. Comeaux personally identified 60 cultivars in Galveston and has recently been studying oleanders in Mexico.

Professor Octavia Hall

Gifted in so many ways, Prof. Hall has led an extraordinary life. A Research Associate Professor with her late husband Dr. Charles E. Hall in the Department of Physiology and Biophysics at the University of Texas Medical Branch in Galveston, Prof. Hall has co-authored 165 research publications. She describes herself as an experimental pathologist and micro-anatomist as well as a string instrument player (mandolin) which, among her many other talents and skills, she credits for contributing to the dexterity necessary for the preparation of herbarium specimens. One set is kept at Moody Gardens and another is at the Smithsonian Institute. In a letter to Prof. Hall, Frederick G. Meyer of the U.S. National Arboretum in Washington, D.C.,

Ted Turner, Sr. with Award from the International Oleander Society

Ted Turner, Sr. and Ted Turner, Jr.

Professor Octavia Hall with her magnificent herbarium sheets of oleanders.

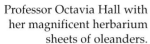

Frank J.J. Pagen

Bob Newding with oleanders, George Sealy (pink), and Sorrento (yellow).

wrote: "I was delighted to receive the specimens of oleander. . . . I must tell you how beautifully they are prepared. All blue ribbon. They will enrich our herbarium of cultivated plants very significantly."

Elizabeth S. Head

A multi-talented member of the International Oleander Society, a medical technologist, writer, poet, painter and native Galvestonian. She served as historian for the Society for many years, is the Corresponding Secretary, the Editor of *Nerium News* and one the Society's strongest pillars. Having grown up on Galveston Island, she has admired oleanders since her childhood and was encouraged in her youth, she recalls, by a science teacher who led the students on field trips and interested them in various horticultural projects. Through the years she has amassed a prodigous collection of data on oleanders compiled into scrapbooks that are an invaluable repository of information on the genus.

Ethyl Mae Koehler

One of the founding members of the National Oleander Society and one of the early presidents, she wrote the first newsletters for the Society, handled all the Society's initial correspondence and authored the Gardening Gal column in the Galveston newspaper.

John Kriegel

Director of Gardens at Moody Gardens, John has been actively involved in landscaping and landscape maintenance for more than twenty years. He has a degree in Botany from the University of Texas, a Masters in Forestry from Stephen F. Austin State University in Texas, and was the Garden Editor for Houston Home and Garden Magazine. At Moody Gardens John also functions in a research capacity in studies of medicinal plants and beneficial insects.

Dr. Darrell McDonald

An ethnobotanist now teaching at Stephen F. Austin State University in Nacogdoches, Texas, his interest in oleanders began in the mid 1980's while a graduate student in geography at Texas A&M University. Fascinated by the exotic mix of plants on Galveston Island, he chose as the subject of his PhD dissertation the evolution and development of the planted landscapes inside the city. His studies made a significant contribution towards a better understanding of the role of oleanders in the flora of the island.

Robert H. Newding

Bob Newding's interest in studying the culture of oleanders and building a representative collection began in earnest about three years ago. He started by asking questions of everyone associated with oleanders, collecting cuttings and propagating hundreds of plants. He now has an impressive collection of more than 70 varieties and, being a trained biologist, has developed an intimate knowledge of oleanders and their culture. The International Oleander Society recently purchased one of each of his plants which are now part of the Society's Reference Collection. Bob has a B.S. in Marine Science from Texas A&M University and is a third generation native Galvestonian.

Gary Outenreath

Gary is the Exhibits Manager at Moody Gardens and one of the most knowledgeable horticulturists in the country. His command of propagation techniques is astounding as is his comprehensive understanding of thousands of rare and unique plants and their culture. He has a B.S. degree in Forestry from Stephen F. Austin State University. He and John Kriegel were the key figures in the landscape design of the Rainforest Pyramid and other plantings throughout Moody Gardens, in the procurement of plants as well as the actual planting and maintenance of the project.

Centennial

F.J.J. Pagen

Frank Pagen studied under Dr. A.J.M. Leeuwenberg at the famous Wageningen Agricultural University in the Netherlands. He is the author of the most authoritative technical treatise on oleanders entitled *Oleanders, Nerium L. and the oleander cultivars*. His extensive research included the examination of hundreds of herbarium specimens from all parts of Europe, especially Mediterranean areas, the Middle East, Africa, India and China, and the review of nursery catalogs from Belgium, France, India, Italy, Japan, Portugal, Spain and the United States, dating back to the early 1820's, from which he compiled a checklist of more than 400 cultivars. He has attended the Oleander Festival in Galveston as a guest lecturer, has generously given of his time to take members of the International Oleander Society to view oleanders in Holland and has recently offered to work with the Nomenclature Committee of the Society where he would no doubt provide invaluable assistance. We applaud his altruistic attitude expressed in a letter to *Nerium News* in which he writes, "It is important to cooperate and share knowledge and information, and to avoid duplication in research." He now has a garden design studio, Ornamental Design, in Spykenisse near Rotterdam.

Dr. Mary Pinkerton

A PhD in botany, a mycologist and the author of many scientific papers, Dr. Pinkerton was a very active member in the early days of the National Oleander Society. She was instrumental in collecting, photographing and classifying the different oleander cultivars in Galveston and helped in setting up the marked oleander trails at SeaArama.

Magnolia Willis Sealy

Wife of George Sealy, Sr., Magnolia is remembered for her work in popularizing oleanders in Galveston. Leader of the Women's Health Protective Association founded in 1896, she spearheaded the organization's drive to rebeautify Galveston after the terrible hurricane of 1900 by planting oleanders in yards and public gardens, and by propagating and distributing plants to residents of the island. By 1912 Mrs. Sealy had developed red and yellow varieties from seed and throughout her life continued to encourage interest and appreciation for the flower she so loved. The variety named after her is considered by all to exemplify the best qualities in an oleander — beautiful, fragrant, long-blooming and hardy.

George Sealy, Jr.

Referred to by his son George Sealy III as "the Johnny Appleseed of the oleander business," he was a pioneer in hybridizing oleanders in Galveston. George Sealy, Jr. made deliberate crosses using hand pollination and named many of his seedlings after prominent men and women of Galveston. With Ed Barr as his right hand man, he propagated hundreds of thousands of plants from seeds and cuttings on the grounds of his cotton concentration company and distributed them free of charge to anyone who agreed to take care of them, including visitors to Galveston and military servicemen who carried his cultivars far and wide. Many of the thousands of plants distributed to residents of Galveston are still gracing the streets and gardens of the island.

George Sealy III and Lane Taylor Sealy

Sons of George Sealy, Jr., their childhood memories include many, often humorous, adventures in connection with their father's passionate hobby, oleanders. (Several of these are recounted in Chapter 11, "Galveston, the Oleander City.") George and Lane continue the Sealy family's long history of philanthropy by supporting various projects in Galveston. In recent years they arranged for the grounds of Open Gates, the former family home which has since been given to the University of Texas Medical School, to be re-landscaped incorporating many of the traditional Galveston oleanders in the design.

Mary Trube

An early member and officer of the International Society, Mary Trube was a musician and philanthropist who, upon her death, left a fund to assist organizations in various fields including music and horticulture.

Ted L. Turner, Sr.

A respected nurseryman and oleander hybridizer, Ted is the recipient of the first Presidential Award from the International Oleander Society for his 'Shari D' cultivar and considers the award one of the great highlights of his years of hybridizing. Ted Sr. has been in the nursery business for more than 40 years, beginning with his father in Premont, Texas. He founded Turner's Gardenland in Corpus Christi twenty-four years ago and has been blessed to have his children and grandchildren grow up around him, learning about the plants and flowers he loves and cherishes.

Ted Turner, Jr.

Though Ted Jr. has followed in his dad's footsteps at Turner's Nursery, he first graduated cum laude in physics and chemistry at the University of Houston (one of only two schools in the United States that offered a computer science degree at that time) where he majored in computer science and mathematics. Having listened to President Kennedy's exhortation to go into the sciences, Ted worked for NASA in Houston throughout the Gemini and the Apollo space missions. His division wrote programs for the onboard computers and the control center for each mission, and then worked in the control center in support of the flight controllers while the missions were in progress. Having grown up in the nursery business, Ted was happy to go back and finds joy in watching his own children grow up in what he calls a "lovely, family oriented business."

Keep company with those who may make you better.
English proverb

Members of the International Oleander Society

(left to right) Betty Lucas, Ken Steblein, Lori Pepper, Clarence Pleasants

(left to right) Elizabeth Head, Sherry Brahm, Oralee Vaughn,
Lynn McNeely, Donna Cox

13

Happiness is not so much in having as sharing.
We make a living by what we get,
but we make a life by what we give.
Norman MacEwan

International Oleander Society

Through the vision and initiative of Clarence Pleasants and Kewpie Gaido, the National Oleander Society was founded in Galveston in May of 1967 (the name was later changed to the International Oleander Society) and the society has been instrumental in popularizing oleanders in America ever since.

During my last visit with Kewpie in June 1995, just months before her untimely passing, she related the story of how the society came into being. After reading a review of Clarence's book in a Houston newspaper, Kewpie was determined to meet him. Since her daughter was attending school in Virginia, she had the opportunity to visit Norfolk where Clarence lived. During their talks he mentioned that El Paso, Texas, wanted to become the Oleander City and while visiting a friend at Longwood Gardens he had learned that the DuPont family might form an Oleander Society. Clarence told Kewpie he knew quite a lot about oleanders and he felt Galveston was truly the Oleander City. He further stated that in all the literature he researched, whenever he found a reference to oleanders, Galveston was usually mentioned. Kewpie gives Clarence full credit for being the inspiration and impetus that sparked the founding of the Society.

The aims of the society are inspiringly set out as follows:

As scientific or horticultural corporation, for non-profit purposes, to promote, carry on and aid in every way the development, improvement and preservation of Oleanders (Nerium) of all kinds, including the importation and improvement by cultivation and hybridization of Oleanders (Nerium): to conduct or cause to be conducted scientific researches for the improvement, development or preservation of Oleanders and for the promotion of a higher degree of efficiency in growing thereof: to collect information relating to the growing and development of Oleanders, to disseminate information concerning the culture, hybridization or development of Oleanders by means of exhibitions, lectures, publications or otherwise, to assist those engaged in the growing of Oleanders by such researches and dissemination: to make awards in the form of certificates, medals or otherwise, for excellence in the development or culture of Oleanders, and generally to extend the knowledge, production, use and appreciation of Oleanders of any kind and in any manner.

According to Clarence Pleasants, Kewpie Gaido, John Kriegel and others who have worked extensively with oleanders in Galveston, not all residents of the island are so enamored of the plant. In a conversation with Prof. Octavia Hall, Pleasants remarked that many in Galveston regard it as "a plebeian plant, common, not with much prestige." Add to this the usual response regarding its poisonous qualities and one can see that the International Oleander Society, through the dedication and persistence of its members, has accomplished a remarkable task in transforming the public image of oleanders and making them one of the most sought after plants for color in the landscape. In 1990 the society was granted tax-exempt status and, with the aid of special grants, has been able to plant hundreds of oleanders throughout the city.

East End Pink

Other noteworthy projects include identification of all varieties found on Galveston Island undertaken by the Nomenclature Committee; building a representative collection for the preservation of named varieties and cultivars from which cuttings and plants can be given to interested gardeners; the production of a video entitled "Galveston, the Oleander City" (Part I has been completed and is available from the society); and regular meetings and workshops to promote knowledge and appreciation of the oleander.

The society is also the official registrant for new cultivars. If you have an oleander that you believe is truly different, the Nomenclature Committee will review your information and evaluate the plant and flower. If they agree that the variety is unique then you have the right to name it and register the name with the society.

One of the most popular projects undertaken by the society is the annual Oleander Festival. On April 28, 1994, the mayor of Galveston proclaimed May as official oleander month for the city. Each year around the third week of May the society presents the Oleander Festival, a two-day event featuring guest speakers, dance, music, decorations and numerous children's events, all highlighting the beauty of oleanders.

The International Oleander Society accepts contributions to further their research on oleanders. They are currently seeking a plot of land on which they can develop a living germ plasm bank by collecting and planting all the named cultivars in one park-like setting. For further information on oleanders and the International Oleander Society write to:

International Oleander Society
P.O. Box 3431
Galveston, TX 77552-0431

Turner's Kathryn Childers™

In every man's heart there is a nerve
that answers to the vibrations of beauty.
Christopher Morley

Turner's Shari D™

14

Among the grasses,
A flower blooms white,
Its name unknown.
Shiki

Plant Patents

The world is indebted to those who are instrumental in creating and/or discovering new plants. We have included this section on plant patents to encourage anyone who has ever considered hybridizing, or who may have discovered a new and unique plant, to understand the United States patent process as given in the examples below. If a new plant is truly of merit, having characteristics superior to the type whether in flower form, shape, color, size, length of flowering season or plant habit, etc., one does not have to be a nurseryman to produce it commercially. There are many growers who will gladly pay a royalty for the opportunity to propagate a new cultivar and make it available to the gardening public.

It is quite fascinating to read about the process of patenting and we have attempted to include all the information you will need to begin. Ted Turner, Sr. says the first and most important step is to find a good patent attorney who is registered with the Patent Office. Formerly, it would take three years or more to get a "Patent Pending." Today, with the aid of computers, it usually takes less than a year.

Turner cautions that one must be very careful that everything is filled out correctly on the application (hence the need for a good patent attorney) because if there are any errors you will have to wait another year before re-applying. If a plant is sold under a "Patent Pending" label one takes a chance that the patent might not clear the Patent Office and during that period anyone can patent the plant. Turner strongly advises that one only patent something truly exceptional.

"HOW TO REGISTER PATENTS, TRADEMARKS"
(from *Nursery Manager*, July 1989)

The Plant Patent Act, passed in 1930, gives certain rights to the discoverer or developer of a plant that is reproduced asexually.

That plant must be new and distinct from any other, and that distinct difference — be it in habit, color, blossom, fruit, fragrance, adaptability, etc. — must be reproducible through asexual propagation.

Such plants may be the result of planned hybridization, seedling selection, naturally occurring mutations, or "sports."

Plants reproduced by seed are protected under the Plant Variety Protection Act; the owner controls multiplication and marketing the protected seed for a period of 18 years.

Plant patent protection lasts for 17 years from the date the patent is granted. Current law requires that a plant patent application contain a varietal or cultivar name. Therein lies some confusion because U.S. law prohibits trademarking of variety (cultivar) names. So a patent owner may choose to market his special plant under a trademark name that must be distinctively different from the cultivar name. Trademarks last for 20 years and are renewable every 20 years so long as the mark is in use.

As of June 6, (1989) some 6840 plants have been patented since 1931, when the first U.S. patent was issued to the rose 'New Dawn.'

The national Association of Plant Patent Owners has a primary purpose "to foster general environmental enhancement by improving varieties of plants for the betterment of man."

The organization monitors federal legislation and regulations that impact plant patents or trademarks, and it works with international organizations of plant breeders on matters of common interests, keeping abreast of laws covering plant protection in foreign nations.

"NAPPO has funded research to provide more scientific means of plant identification through the use of plant fingerprinting," said Ben Bolusky, administrator of NAPPO and director of governmental affairs for the American Association of Nurserymen.

The International Convention for the Protection of New Varieties of Plants, signed by the United States and 16 countries, currently is undergoing wholesale revision in meetings in Geneva, Switzerland. NAPPO has been asked by the federal government to attend.

"The U.S. is the only nation that allows such private sector advisers to the delegation," Bolusky said.

For more information about books and lists, membership in NAPPO, or member law firms, contact the National Association of Plant Patent Owners.

"How to Use, Select and Register Cultivar Names," available from the AAN may also be useful.

Contact both NAPPO and AAN at 1250 I St. N.W., Suite 500, Washington, D.C. 20005; (202) 789-2900; fax (202) 789-1893.

Everything should be made as simple as possible, but not simpler.
Albert Einstein

Glossary

Anther: The pollen-bearing part of the stamen.

Apical fringe: The top portion of the corona which is usually dissected into a fringe.

Bicarpellate: Composed of two carpels. A carpel is one of the units that compose a pistil or ovary.

Bract: A modified leaf, usually smaller than true leaves and associated with the flowers. They may be colorful and showy as in poinsettias and bougainvilleas.

Calyx: The outer whorl of floral envelopes, composed of separate or united sepals.

Campanulate: Bell-shaped.

Caudex: The swollen stem base of certain plants.

Coma: A tuft of soft hairs on a seed.

Corolla: The inner circle or whorl of floral envelopes. A corolla is the colored, showy part of the flower consisting of fused petals and corona. When the parts of the floral envelope are separate they are called petals.

Corona: Literally, a crown. A circular petal-like appendage in the center of the corolla.

Cultivar: A horticultural variety that originated under cultivation and is of sufficient importance to require a name. Cultivar names follow the genus and species and are indicated by the use of single quotation marks.

Cuneate: Basically wedge-shaped or narrowly triangular, with the narrow end at the point of attachment.

Cyme: A type of inflorescence characterized by being determinate and broad, in which the central flowers usually are the first to open.

Dodder: A parasitic climber that attaches itself to a host plant.

Exserted: Projecting out.

Flocculation: The process of neutralizing electrical charges on colloids. In clay soils the addition of gypsum and compost or other organic matter is generally utilized, allowing the minute particles of clay to adhere to other particles, thereby forming larger aggregates insuring that oxygen and drainage will be present to allow root penetration.

Follicle: A dry, dehiscent fruit with usually more than one seed and opening only along the ventral suture.

Gall: Gall may appear on any part of a plant and has many causes, including fungi, bacteria, viruses, nematodes and insects. Evidence of gall infestation is usually a tumor or growth in a wide diversity of shapes.

Glabrous: Smooth, without hairs of any kind.

Glycosides: Sugar derivatives widely found in plants.

Half-hardy: Somewhat resistant to damage from freezing.

Inflorescence: A flower cluster having a common axis.

Introrse: Facing towards the center or inward; as an anther that dehisces towards the center of a flower.

Nectaries: The nectar-secreting glands.

Petaloid: To resemble a petal, or a part of a plant that resembles a petal.

Petiole: A leaf stalk.

Pistil: The seed-bearing part of a plant consisting of the stigma, style and ovary.

Puberulous: Covered with hairs that are minute, soft and erect.

Rank: A vertical row. When a flower is single, it has one row of petals, or one rank. Two ranks indicate a double flower.

Sagittate: Shaped like an arrowhead. In leaves, with the basal lobes pointing downward.

Sepal: One of the separate parts of a calyx, usually green.

Sport: A plant or part of a plant exhibiting marked variation from normal; a mutation.

Stamen: The pollen bearing organ which consists of the anther and filament (stalk).

Stigma: The tip of the pistil upon which the pollen falls and germinates.

Style: The usually elongated part of the pistil between the stigma and ovary.

Taxa: Scientific classifications within a system.

Taxonomist: A person specializing in the science of classification of objects, animals or plants.

Leisure, slowness, contemplation: in an age of presumed efficiency and professionalism, these amateur virtues are perhaps despised, but they may underlie the greatest joys of gardening, and of life. It is not enough to grow the most beautiful things. It is even better to explore them, to identify with them, and to grow into a rather new consciousness of them...

Henry Mitchell,
The Essential Earthman

BIBLIOGRAPHY

Abbott, Daisy T., *The Indoor Gardener*, Univ. of Minnesota Press, Minneapolis, MN, 1939.

Adams, William D., *Shrubs and Vines for Southern Landscapes*, Gulf Publishing Company, Houston, TX, 1979.

Arp, Gerald K., *Tropical Gardening Along the Gulf Coast*, Gulf Publishing Company, Houston, TX, 1978.

Baines, Jocelyn and Katherine Key, *The ABC of Indoor Plants*, Alfred A. Knopf, New York, NY, 1973.

Barrick, William E., *Salt Tolerant Plants For Florida Landscapes*, Univ. of Florida, Gainesville, FL, 1979.

Bateson, Wade T., *Landscape Plants for the Southeast*, Univ. of South Carolina Press, Columbia, SC, 1984.

Beckett, Kenneth A., *The Royal Horticultural Society's Encyclopedia of House Plants*, Salem House Publishing, Boston, MA, 1987.

Bender, Steve, "Oleander Deserves a Second Chance", Southern Living, June 1995.

Bird, Richard, *Flowering Trees and Shrubs*, Quarto Publishing, London, 1989.

Boyd, Lizzie, *Window Gardens*, Clarkson N. Potter, Inc., New York, NY, 1985.

Bridwell, Ferrell M., *Landscape Plants*, Delmar Publishers, Albany, NY, 1994.

Bruggeman, L., *Tropical Plants and Their Cultivation*, Thames and Hudson, London, England, 1962.

Chittenden, Fred J., *Dictionary of Gardening*, The Royal Horticultural Society, Oxford at the Clarendon Press, England, 1951.

Coats, Alice M., *Garden Shrubs and their Histories*, Simon and Shuster, New York, NY, 1992.

Corpus Christi Area Garden Council, Inc., *The Garden Book*, Golden Banner Press, 1992.

Courtright, Gordon, *Trees and Shrubs for Temperate Climates*, 3rd ed., Timber Press, Portland, OR, 1979.

Courtright, Gordon, *Tropicals*, Timber Press, Portland, OR, 1988.

Evans, Charles M., *New Plants from Old*, Random House, New York, NY, 1976.

Everett, Thomas H., *The New York Botanical Garden Illustrated Encyclopedia of Horticulture*, Garland Publishing, Inc., New York, NY, 1981.

Evergreen Shrubs, The Time-Life Gardener's Guide, Time-Life Books, Alexandria, VA, 1989.

Fancher Creek Nursery Catalog, Fresno, CA, 1905.

Flint, Harrison L., *Landscape Plants for Eastern North America*, John Wiley & Sons, New York, NY, 1983.

Flowering Houseplants, The Time-Life Gardener's Guide, Time-Life Books, Alexandria, VA, 1990.

Foster, Raymond, *Rare and Exotic Plants*, The Overlook Press, Woodstock, NY, 1985.

Free, Montague, *All About House Plants*, Doubleday & Company, Garden City, NY, 1979.

Free, Montague, *Plant Propagation in Pictures*, The American Garden Guild, Inc., and Doubleday & Company, Inc., Garden City, NY, 1957.

Gopalaswamiengar, K.S., *Complete Gardening In India*, 4th ed., Rev. and pub. by Gopalaswamy Parthasarathy, Bangalore, India, 1991.

Graf, A.B., *Tropica*, 3rd ed., Roehrs Company, East Rutherford, NJ, 1986.

Greenhouse Gardening, A Redefinition Book, Time-Life Publishing, Alexandria, VA, 1989.

Halfacre, R. Gordon, *Carolina Landscape Plants*, Ed. Anne Rogers Shawcroft, The Sparks Press, Raleigh, NC, 1971.

Hall, Professor Octavia, Memoirs and personal correspondence to the authors, unpublished.

Hargreaves, Dorothy and Bob, *Hawaii Blossoms*, Hargreaves Company, Kailua, HI, 1958.

Harrison, Richmond E., *Trees and Shrubs*, Timber Press, Portland, OR, 1983.

Hay, Roy and Patrick M. Synge, *The Colour Dictionary of Garden Plants*, Bloomsbury Books, London, England.

Head, Elizabeth S., *Nerium News*, The Newsletter of the International Oleander Society, Galveston, TX.

Head, Elizabeth S., "Oleanders", *American Horticulturist*, American Horticultural Society, October 1985.

Head, Elizabeth S., Personal correspondence to the authors, unpublished.

Herwig, Rob, *2850 House & Garden Plants*, Crescent Books, New York, NY, 1985.

Hillier & Sons, *The Hillier Manual of Trees & Shrubs*, 6th ed., Hillier Nurseries (Winchester) Ltd., Great Britain, 1992.

Hirsch, Doris F., *Indoor Plants*, Chilton Book Company, Radnor, PA, 1977.

Holttum, R.E. and Ivan Enoch, *Gardening in the Tropics*, Times Editions, Singapore, 1953.

Hudak, Joseph, *Shrubs in the Landscape*, McGraw-Hill Book Company, New York,NY, 1984.

International Oleander Society, *Oleanders, Guide to Culture and Selected Varieties on Galveston Island*, Galveston, TX, 1991.

Kirchem, C., Personal correspondence from R.C. Aldridge, Jr., Pres. Aldridge Nursery, Inc., Von Ormy, TX, 1978.

Kriegel, John and Editors of Houston Home & Garden Magazine, *Houston Garden Book*, Shearer Publishing, 1983.

Kromdijk, G., *200 House Plants In Color*, Herder and Herder, New York, NY, 1973.

Krüssman, Gerd, *Manual of Cultivated Broad-leaved Trees & Shrubs*, Vol. II, Timber Press, OR, 1986.

Kuck, Loraine E. and Richard C. Tongg, *Hawaiian Flowers & Flowering Trees*, Charles E. Tuttle Co., Rutland, VT and Tokyo, Japan, 1958.

L.H. Bailey Hortorium, (Staff), *Hortus Third*, Cornell Univ., Macmillan Publishing Company, New York, NY, 1976.

Lunardi, Costanza, *Guide to Shrubs and Vines*, Simon and Schuster, New York, NY, 1988.

Macoboy, Stirling, *What Shrub is That?*, Portland House, New York, NY, 1990.

Martin, Edward C., Jr., *Landscape Plants in Design*, AVI Publishing Co., Westport, CT, 1983.

Martin, Tovah, *Once Upon A Windowsill*, Timber Press, Portland, OR.

McDonald, Elvin, *The New Houseplant*, Macmillan Publishing Co., New York, NY, 1993.

Morton, Julia F., *Exotic Plants*, Golden Press, New York, NY, 1971.

National Gardening Association, The, *Dictionary of Horticulture*, Penguin Group, New York, NY, 1994.

Neal, Marie C., *In Gardens of Hawaii*, Bishop Museum Press, Honolulu, HI, 1965.

New Royal Horticultural Society Dictionary of Gardening, (The), 3 vols., Ed. Anthony Huxley, Mark Griffiths and Margot Levy, Macmillan Press Ltd., London and Stockton Press, New York, NY, 1992.

New Western Garden Book, 4th ed., Ed. Sunset Books and Sunset Magazine, Lane Publishing Co., Menlo Park, CA, 1967.

Nursery Manager, "How to Register Patents, Trademarks", Branch-Smith Publishing, Fort Worth, TX, July 1989.

Odenwald, Neil, and James Turner, *Southern Plants*, Claitor's Publishing Division, Baton Rouge, LA, 1987.

Pagen, F.J.J., *Oleanders, Nerium L. and the oleander cultivars*, Agricultural Univ. Wageningen Papers, The Netherlands, 1987.

Pleasants, Clarence G., *Galveston, The Oleander City*, Exposition Press, New York, NY, 1966.

Pleasants, Clarence G., *Oleanders*, Pamphlet, March 1967.

Pleasants, Clarence G., Personal correspondence from George C. Roeding, Jr., California Nursery Company, Fremont, CA, 1981.

Pleasants, Clarence G., Personal correspondence from Donald J. Moore, Superintendent, Botanical Gardens, Bermuda, 1964.

Pleasants, Clarence G., "The Spectacular Oleander", *Horticulture Magazine*, Boston, MA, 1969.

Randhawa, M.S., *Flowering Trees in India*, Indian Council of Agricultural Research, New Delhi, India, 1957.

Schuler, Stanley, *Basic Book of Trees and Shrubs*, Simon and Schuster, New York, NY, 1973.

Seddon, George and Andrew Bicknell, *Conservatory Gardening*, Willow Books, Collins, London, 1986.

Seiden, Allan, *Flowers of Hawaii*, Island Heritage Publishers, 6th ed., Honolulu, HI, 1988.

Smiley, Nixon, *Tropical Planting and Gardening*, University of Miami Press, Coral Gables, FL, 1960.

Southern Living, "Oleander Deserves a Second Chance," June 1995.

Sri Aurobindo Society, *Flowers — their Spiritual Significance*, 2nd imp., Ed. Vijay, Sri Aurobindo Ashram Press, Pondicherry, India, 1993.

Stresau, Frederick B., *Florida, My Eden*, Florida Classics Library, Port Salerno, FL, 1986.

Taylor, Jane, *Drought-Tolerant Plants, a Horticulture Book*, Prentiss Hall, New York,NY, 1993.

Taylor's Guide to Container Gardening, Ed. Roger Holmes, Houghton Mifflin Company, Boston, MA and New York, NY, 1995.

Taylor's Guide to Shrubs, Houghton Mifflin Company, Boston, MA, 1987.

A Garden Book For Houston, Ed. Lorna Hume Terrell, 4th ed., The River Oaks Garden Club, Houston, TX, 1989.

The Mother, *Flowers and Their Messages*, Sri Aurobindo Ashram, Pondicherry, India, 1979.

Toogood, Alan, *The Hillier Guide to Connoisseur's Plants*, David & Charles, Newton Abbot, Devon, England, 1991.

Turner, Sr., Ted, "Nerium Oleanders," Address to the annual meeting of the International Oleander Society, unpublished, 1988.

Watkins, John V. and Thomas J. Sheehan, *Florida Landscape Plants*, Univ. of Florida, 1975.

Whitcomb, Carl E., *Know It and Grow It, A Guide to the Identification and Use of Landscape Plants*, Lacebark Publications, Stillwater, OK, 1983.

Wiggington, Brooks E., *Trees and Shrubs for the Southern Coastal Plain*, Univ. of Georgia Press, Athens, GA, 1957.

Williams, Winston, *Florida's Fabulous Flowers*, World-Wide Printing, Tampa, FL, 1986.

Wyman, Donald, *Hedges Screens & Windbreaks*, Whittlesey House, a div. of McGraw-Hill, New York, NY, 1938.

Wyman, Donald, *Shrubs and Vines for American Gardens*, Macmillan Publishing Company, New York, NY, 1969.

Zafrir, D., *The Nerium Oleander in Israel*, trans. A. Gorvitz, Hakibutz Hameuchad, Pub., Israel, 1962.

Hawaii

154

INDEX OF COLOR PHOTOGRAPHS

Illustrations

Please note that the ISBN number in the front of this book on page iv is incorrect.
The correct ISBN number for this book is
ISBN 0-9643224-1-2